CONCISE OF
CHEMICAL
ELEMENTS

CONCISE OF CHEMICAL ELEMENTS

Jitendra Kumar Vimal
M. Tech.

Lecturer (Civil Engineering)
Govt. Polytechnic, Dhanbad

White Falcon
Publishing

www.whitefalconpublishing.com

Concise of Chemical Elements
Jitendra Kumar Vimal

www.whitefalconpublishing.com

ISBN - 978-93-89530-27-8

Preface to the First Edition

A branch of Science concerned with the properties, composition of substances and their reactions with one another is known as chemistry. I felt the need for a book on "Concise of Chemical Elements" for chemistry students. There are several books on this subject but differences run deep. I had decided to accept this work in spite of my hard- work schedule.

I am confident that this book would not only serve the needs of the student community at large but would also stimulate their interest in the discovery of chemical elements.

It is more intimately connected with our daily life. The food we eat, the clothes we wear, the building materials we use, the medicines we need to maintain health and other human necessities are all dependent on the chemical knowledge. Tremendous developments in science and technology would not have taken place without the development in chemistry. The advancement in communication, space travel, agriculture and so many other fields have been made possible mainly due to the developments in chemical elements.

Starting from basic concepts and gradually going into the required depth. The subject matter is presented in such a way that students do not feel any difficulty in understanding the matter and should be able to grasp the various aspects of chemistry. It is hoped that the present book will open up new vistas of chemistry to an average student and create in him the curiosity to acquire more knowledge and to gain sufficient understanding of the discovery of chemical elements.

My sincere thanks are due to all my seniors and colleagues who encouraged me to write this title. I extend my special thanks to all my friends and well-wishers who have co-operated directly or indirectly in the completion of the task. I am thankful to all my past and present students who have insisted me to accept this work.

I express my heartiest thanks to Mr. Ajit John Barla, for his constant inspiration, deep concern, co-operation and encouragement throughout the project. I cannot forget how he helped in every way; he keeps a special place in my heart.

I extend my heartfelt thanks to Dr. Laldeep Gope, Senior Technical Assistant., Department of Petroleum Engg. IIT (ISM), Dhanbad for giving valued suggestions for a better presentation of this book.

I wish to express special thanks to Prof. Ajay Kumar (Head of the Dept. of Civil Engg. Govt. Polytechnic, Nirsa) for his tireless efforts during the processing of the manuscript.

I express my deep sense of gratitude and indebtedness to Prof. K.P. Yadav (Ex Head of the Dept. of Civil Engg., Govt. Polytechnic, Dhanbad) for his invaluable advice and suggestions from time to time.

I shall be deeply grateful to readers who will give me their frank opinion about the book and make suggestions for further enrichment of this book. Any suggestion is welcome, useful suggestions to make these books better and better still.

I hope my efforts will be useful to chemistry students. I feel the students, as well as the teachers, will benefit a lot from this book since the matter is very simplified for understanding.

Enjoy reading
Well Wisher

Er. J.K. Vimal

A few words about the structure of the Book

It gives me enormous pleasure to introduce "Concise of Chemical Elements". I have tried my best by putting forth my lifetime's experiences as a teacher into this book. I am sure the book will be extremely well-received by a large community of students, teachers, the common man and even scientists. Reader's Satisfaction, Author's success.

It consists of two parts. The first part deals with the natural elements, the second part - with the synthesized ones.

Concise of Chemical Elements has been written concisely and divided into 14 chapters according to type of elements.

How these elements were discovered is the subject of the first part of our book.

How the inquisitive mind of the researchers discovered new chemical elements, one after another.

Each chapter carries important information about the topics so that students gain more knowledge about the given topics in all chapters.

Each chapter contained in the book consists of a detailed treatment of the subject matter and is written in a very simple language which is easy to read, user-friendly and easy to understand.

Each element has its biography, interesting in its own way. Hence, the depth of knowledge of the subject gained will help the students to have continuous access to these topics and to achieve excellence in the path of the quest for chemical element knowledge.

Every line of the book is pact with important, essential information that it provides interesting reading material, more enjoyable on some lesser-known aspects, which makes it a valuable contribution.

Each chapter to help the readers to have a bird's eye view of the chapter's content.

Each element has an important role to play in today's advanced science and technology.

The book discusses the elements found in the atmosphere and the earth's crust.

The book is all about chemical elements and the vital role they play in our lives.

The book has been written about practically all chemical elements - enough to stock a great library.

The book is written to meet certain specific objectives. One of these objectives is that students should be able to appreciate methods of science, measurements in chemistry, make correct observations, arrange facts and draw valid conclusions.

Acknowledgement

I special acknowledge the professional staff at White Falcon for their efforts in bringing this project to completion. Ms Navsangeet Kaur (Director); She is extremely Sweet and dynamic. She has been the backbone behind publishing this book. we are gald to work with you mam. We Salute them, for their professionalism. I express my heartfelt thanks to Ms Gunjan Rastogi, for their constant inspiration, deep concern, cooperation and encouragement throughout the project.

I heartily thank my dear friend Arti Munda, B.Tech. (Computer Science) for her selfless and prompt help in completion of this book. I cannot forget how they have helped in every small or big way, she has a special place in my heart.

I am also especially thankfull to Tanya Gupta for her active interest, for giving me boost every time, for her best wishes, for giving me moral support to write.

I DEDICATE THIS BOOK TO

My Grandfather Late Barhu Mahasya
My Grandmother Late Sundari Devi
My mother Late Uma Devi
and
My father Gyan Prakash Vimal

Our Parents for their abundant support, their patience and understanding, their sacrifices and for their love. Whose love and energy make everything I do possible. Whose blessings and constant encouragement have always been with me throughout my endeavour.

Author Biography

I, Jitendra Kumar Vimal, S/o Shri Gyan Prakash Vimal, solemnly affirm and declare that I am 49 years old energetic man born on 1st January 1969 in the village Tungi, Biharsharif belonging to the district Nalanda, State Bihar (India). I was brought up in Patna, a prominent Capital city of Bihar, where my father was serving in Central Government job.

I got the degree of M. Tech in the year 1999. Presently, I am serving as a teaching faculty at the Govt. Polytechnic, Dhanbad, for over one decade as a Lecturer of Civil Engineering.

I write books on engineering and science subjects, poems and national issues in simple language given the present situation and demand.

Below mentioned books are going to be completed in near future. These books provide interesting reading material and in-depth understanding of the subject, which is user-friendly and easy to understand.

1. Concise of Periodic Table
2 Concise of Physics
3. Concise of Chemistry
4. Concise of Biology

5. Concise of Construction Material
6. Concise of Concrete Technology
8. Science of Chronology
9. Science of Knowledge
10. Science of Planetary
11. Soil Mechanics
12. Highway Engineering
13. Kavita Sangrah
14. Desh ka Durbhagya
15. Janta Ki Durgati Akhir Kyun?
16. Daastan-E-Ambedkar
17. Long form of the Modern Periodic Table chart
18. Andaaz-E-Ishq

I desire to contribute 10% of my royalty to the World Health Organization (W.H.O.) for humanitarian purposes. I have thought so because I have spent my life in dearth of means to lead a purposeful life.

I am also eager to expend 10% of my royalty (besides 10% contribution to W.H.O.) in India to improve the level of literacy, health, hygiene and in awareness program for those who are living below the poverty line or for down-trodden and marginalized people of India.

I think that my contribution to the W.H.O. from my royalty will be much more helpful for the needy class of the world society, who are spending and leading their sub-standard and sub-human life worldwide.

I think that it will be a great opportunity for me to serve the poor people who are suffering from lot of problems, malnutrition and several diseases in the world.

I believe that book writing is just like classroom teaching as I enjoy and feel proud to contribute my knowledge through my writing to the world society.

I think that every writer has his view in writing on any aspect and that writing on particular aspects and fields will be helpful for the society.

I think that every reader has his view and thinking on any book or writing. Readers generally like those books that are more suitable and sufficient for their course. But sometimes, besides the course book, some general books will also be much more helpful in increasing their mental level. Such books may change their attitude, habits etc. and spread good knowledge in their life.

Thus, readers never say that a particular book is not useful. That book may be useful for others. Thus the readers should be broad in their mind, attitude and thinking.

I like to work in a challenging environment where my skills and knowledge can be utilized in the best interest of the organization and help achieve both professional and personal goals with all my endeavors.

I am a work-oriented energetic person, a voracious learner who never claims to be perfect but strives to be a perfect learner.

Some of my strong points are:-

- During my working hours, I always try my best to satisfy my seniors as well as my juniors and the students of my class by dint of my hard labor, honesty, punctuality, and sincerity.
- Interested in learning new things, ability to work under stressful conditions
- Language well known - English, Hindi, and Russian
- Hard work can breakdown every obstacle
- Honesty is the best policy. Honesty, as well as hard work, are the best way to succeed in life.

Some inspiring messages for students:-

- God has given us two ears and two eyes but only one tongue so that you can hear and see more than you speak.
- Never postpone a thing for tomorrow if it is possible for you to do it today.
- Never hurt anybody. Control anger by love, forgiveness and compassion.

- Reduce your wants, lead a happy, contented life, avoid unnecessary worry.
- Develop virtues:- Forgiveness, mercy, love, perseverance, truthfulness, patience, kindness, courage, etc.
- Don't think about what can happen in a month. Just focus on the 24 hours in front of you and your success is hidden in your daily routine.
- To make a nation great, three qualities are needed. "Nobility in leadership, indomitable spirit and universal mind." These will be utilized to convert dreams to thoughts and finally lead these thoughts to action.
- I wish you all the successes in your future. I wish you all the success of your career.

Special importance of books in human life:-
- Books are the source of infinite wisdom. The more they are studied, the more you get the knowledge from them.
- Books, where we are given knowledge, are the companions of the teacher, friend, associate, path - shower and long travel.
- Books teach us to love our traditions and culture. This makes it worthwhile to understand the pain of others and to make a better contribution to society.
- If you are thirsty, dig a well and drink water. Wake up to reading and read new books.
- Books float from the mirrors of closed wardrobe. A house without a book is, like a house without windows.
- New books are now just like the brides coming out of the *doli*, they look bright and dazzling. Old books say that we have not been touched till date, then why do we bring new books? How to explain them, even if they are not read, their presence gives feeling of peacefulness.
- A room without a book is like a body devoid of soul. For today and forever, the best friend is good books.

- Life is a fair and books are required to understand that fair. Books are among the coolest and lasting friends. The advisors and the teachers are the most patient teachers.

Some important tips which will help you in setting your goals in studies:-

1. Set Goals That Motivate You: This means making sure that they are important to you and that there is value in achieving them.

2. Set SMART Goals:

   Specific: Your goal must be clear and well defined, not vague or generalized.

   Measurable: Goals must have measurable objectives.

   Attainable: Make sure your goals are achievable and within your limit.

   Relevant: Will take you to the direction you want your life and career to go.

   Time-Bound: You must know when you have the deadline and can celebrate success.

3. Set Goals in Writing: Written commitment in the presence of your close people (parents, close friends) will always push and remind you whenever you tend to deviate from your goal.

4. Make an Action Plan: Do not focus only on the outcome but make planning of all small steps that collectively take to the outcome. This is especially important if your goal is big and demanding or long-term.

5. Monitor Yourself: Compliance to the action plan should be monitored atleast weekly (for a one-month goal) or monthly (for a yearly goal), depending upon your goal size.

Remember,

"Success is not final; failure is not fatal: it is the courage to continue that counts."

"There are two types of people who will tell you that you cannot make a difference in this world: those who are afraid to try and those who are afraid you will succeed."

"Success ke piche mat bhago, kabil bano kabil. Kamyabi to Sali jhak maar ke piche ayegi."

1. पुस्तकें अनंत ज्ञान का स्रोत है। इनका जितना ही अध्ययन किया जाए, उतना ही उनसे ज्ञान की प्राप्ति होती है।

2. पुस्तकें जहां हमें ज्ञान देती हैं वही वह गुरु, मित्र, सहयोगी, पथ – प्रदर्शक और लम्बी यात्रा की संगिनी होती है।

3. पुस्तकें हमें अपनी परम्पराओं और संस्कृति से प्रेम करना सीखती है। इस काबिल बनाती है की दुसरो के दुःख दर्द को समझ सके और समाज में कुछ बेहतर योगदान दे सकें।

4. प्यास लगे तो कुआँ खोदकर पानी पीओ। पढ़ने की प्यास जगाओ और नयी – नयी पुस्तकें पढों।

5. नई किताबें अभी – अभी डोली से उतरी दुल्हन की तरह चमक और दर्प से भरी दिखती है। पुरानी किताबें उलाहना देती है की आज तक हमें ही नहीं छुआ, तो नयी किताबें क्यों लाए? उन्हें कैसे समझाऊं कि नहीं पढ़े जाने पर भी उनकी मौजूदगी कितना सुकून देती है।

6. किताबें झांकती है, बंद अलमारी के शीशो से। बिना किताबो के घर उसी तरह है, जिस तरह बिना खिड़कियों के मकान।

7. पुस्तक से रहित कमरा आत्मा से रहित शरीर के समान है। आज के लिए और सदा के लिए सबसे बड़ा मित्र है अच्छी पुस्तकें।

8. जिंदगी एक मेला है और किताबें उस मेले को समझने का हथियार। किताबें मित्रों में सबसे शांत और स्थायी मित्र होती है। सलाहकारों में सबसे सुलभ और बुद्धिमान सलाहकार होती है, और शिक्षको में सबसे धैर्यवान शिक्षक होती हैं।

A Word to the Readers

"Chemistry" is derived from the word "kheem" which is the old name of Egypt and probably given to it due to the black colour of the Egyptian soil. The art of khemia flourished in the early Egyptian and Greek Civilizations. The word khemia then became Al-kimiya in Arabic and English word - Alchemy was later derived.

The language of chemistry has its own alphabets. Its letters are symbols of chemical elements; the number of combinations of letters, words composed of them, is infinite – the endless variety of chemical compounds. More than four million chemical compounds are known at present and this number increases each week by six thousand. This "word–building" in chemistry is a non–stop process.

Individual letters or elements are much fewer in number. Today, there are only one hundred and eighteen of them. Several thousand years were required to compile the alphabets of the language of chemistry but most of the letters were deciphered only during the last two hundred years. It was during this short period that chemistry emerged as a science.

All chemical compounds that constitute living and inorganic matter are diverse combinations of eighty-odd elements. The remaining known elements are practically not found in nature. Scientists created them artificially by means of nuclear reactions. More new elements can be obtained in this manner and we do not know how many of them. But it is quite clear that the chemical alphabets are not complete yet.

In this book, we shall describe how the alphabets of chemistry have been designed and how the inquisitive mind of the researchers discovered new chemical elements, one after another.

The book has been written about practically all chemical elements – enough to stock a great library. They describe minerals and ores containing chemical elements, processes of their extraction, physical and chemical properties of the elements and their uses. Many elements are surprisingly abundant: they can be used in widely disparate and unexpected fields for the good of mankind. Almost every element has an important role to play in today's advanced science and technology.

The history of chemical elements begins with their discovery. Although hefty volumes in which elements are described in detail pay very little attention to their discoveries, they are a major part of the history of human knowledge.

Each element has its own "biography" interesting in its own way. The history of the discovery of many elements has not yet been exhaustively studied and quite a number of unclear issues should be cleared by historians of chemistry; perhaps you will be one of them.

Contents

Introduction xxvii

Part - I 1
Elements discovered in nature 1

1   Elements known in Antiquity 3
    Carbon 4
    Sulphur 7
    Gold 8
    Silver 11
    Copper 13
    Iron 15
    Lead 17
    Tin 18
    Mercury 20

2   Elements discovered in the Middle Ages 23
    Phosphorus 24
    Arsenic 27
    Antimony 28
    Bismuth 29
    Zinc 30

3 **Elements of Air and Water** 33
   Hydrogen 37
   Nitrogen 42
   Oxygen 46

4 **Elements discovered by Chemical Analysis** 54
   Cobalt 54
   Nickel 56
   Manganese 58
   Barium 60
   Molybdenum 61
   Tungsten 63
   Tellurium 64
   Strontium 66
   Zirconium 68
   Uranium 69
   Titanium 73
   Chromium 75
   Beryllium 77
   Niobium and Tantalum 81
   Platinum Metals 84
   Platinum 85
   Palladium 87
   Rhodium 88
   Osmium and Iridium 89
   Ruthenium 92
   Halogens 94
   Fluorine 94
   Chlorine 99
   Iodine 103
   Bromine 105
   Significance of Halogens for the Development of Chemistry 108
   Boron 109
   Cadmium 110
   Lithium 112
   Selenium 114
   Silicon 116

Aluminium.. 118
Thorium... 122
Vanadium... 123

5   Elements discovered by the Electrochemical Method 127
Sodium and Potassium... 128
Magnesium 131
Calcium 132

6   Elements discovered by the Spectroscopic Method................ 135
Caesium 136
Rubidium... 138
Thallium 139
Indium 142

7   Rare Earths... 145
REEs Early History.. 146
Lanthanum and Didymium, Terbium and Erbium
"Ytterbium", Scandium, "Holmium", Thulium 150
The End of "Didymium", "Samarium", Neodymium
and Praseodymium... 153
Gadolinium and Dysprosium 157
"Time of Confusion" in the History of REEs........................ 158
Ytterbium and Lutetium 161
Lessons of REEs History.. 162

8   Helium and other Inert Gases 164
Helium... 165
Argon... 170
Krypton, Neon and Xenon 175
Inert Gases as Food for Thought 179

9   Elements predicted from the Periodic System.................... 181
Gallium.. 184
Scandium 187
Germanium.. 189
Prediction of Unknown Chemical Elements 192

10 Hafnium and Rhenium–two stable elements
   which were the last to be discovered 195
   Hafnium 196
   Rhenium 199

11 Radioactive Elements 205
   Polonium 206
   Radium 210
   Actinium 213
   Radon 216
   Protactinium 226
   Francium 230

PART - II 233
Synthesized Elements 233

12 Discoveries of synthesized elements within
   the old boundaries of the periodic system 237
   Technetium 237
   Promethium 247
   Astatine and Francium 258

13 Transuranium Elements 269
   Neptunium 277
   Plutonium 280
   Americium and Curium 282
   Berkelium 285
   Californium 286
   Einsteinium and Fermium 287
   Mendelevium 289
   Nobelium 293
   Lawrencium 294

14 Discoveries of Elements 296

Introduction

About eighty years ago Clemens Winkler, the German chemist who discovered germanium which had been predicted by D. Mendeleev under the name of "eka-silicon", likened the world of elements to the theatre stage where scene after scene is played out with elements as characters. Each element, Winkler said, plays its own role. Sometimes it is a subsidiary role; sometimes it is a leading role.

In this way, the Scientist characterized the significance of the elements already discovered and known to man.

From the standpoint of the history of discovery, there can be neither leading nor subsidiary elements. All elements can lay equal claim to our attention.

Therefore, it is up to us to decide in what sequence the history of the discovery of the elements should be presented.

We can describe elements in the order of increasing atomic numbers: hydrogen, helium, lithium... up to element No. 118, which is still. Or we may describe the history of the discovery of the elements that compose the successive groups of the periodic system. Or we may deal with the elements in alphabetical order.

We believe that all these ways of presentation are not very successful since they distort the chronology of discoveries. And it is exactly the chronology that we want to make the basis of the presentation here.

But at first, let us try to understand what is meant by the term "a chemical element".

THE CONCEPT OF A "CHEMICAL ELEMENT"

An element is the totality of atoms of a certain type. An atom consists of a nucleus and electrons surrounding it. A nucleus has an integral positive charge denoted by the Latin letter Z. The charge, in its turn, is determined by the number of elementary particles (protons) contained in the nucleus. The charge of the proton (positive) is equal in magnitude to that of the electron (negative). This means that the number of protons (z) in the nucleus determines the number of electrons in electron shells of the atom. The chemical properties and behaviour of the elements depend on how the electrons are distributed in the shells. Consequently, the nuclear charge Z determines the properties of the chemical element. Also, Z coincides with the atomic number of the element in the periodic table. For instance, the nucleus of the oxygen atom (atomic number 8) has a positive charge equal to 8, i.e., it contains 8 protons.

Thus, an element is a set of atoms with the same nuclear charge Z which determines the position of the element in the periodic system.

Can atoms of the same element differ from one another? The answer proves to be "yes". In addition to protons, a nucleus contains neutrons. As regards their mass, neutrons differ only slightly from protons, but, in contrast to protons, they carry no charge: they are neutral. There are no nuclei without neutrons (the only exception is the nucleus of the lightest element, hydrogen, which is just a single proton; however, there are different types of hydrogen atoms whose nuclei contain neutrons as well). The total mass of protons and neutrons in a nucleus determines the mass of the atom since the masses of electrons are negligibly small (an electron is 1840 times lighter than a proton). The varieties of the atoms of this or that element whose nuclei contain a different number of neutrons are called isotopic atoms or isotopes. The word "isotope" originates from the Greek isos, "the same", and topos, "place". This means that all the isotopes of the same element occupy the same position in the periodic table. About three-fourths of the naturally occurring elements have isotopes or, as is said, represent a pleiad of isotopes. The remaining elements have no isotopes, i.e., they exist only in one variety of atoms.

Even though the concept of a "chemical element" seems to be quite definite, in reality, it is a rather abstract term denoting only a group of atoms with a given nuclear charge. In practice, we deal with elements either as constituents of various chemical compounds or as simple substances. A simple substance is a free form of an element which makes it possible to see what the element looks like. Some elements occur in nature only as simple substances, others – as simple substances or as constituents of compounds and still others exclusively in combinations with other elements. The representatives of the last group are especially numerous. The forms of existence of elements in nature played an important role in the history of their discovery.

WHERE THE NAME "ELEMENT" CAME FROM

Historians of chemistry have no consensus on this question and only more or less plausible assumptions can be made. The fact is that the concept of an "element" used in ancient times was wider in its meaning than that assigned to a chemical element how. It was to a great extent of a philosophical nature.

One of the hypotheses explaining this is as follows. The word "element" originates from the letters of the Latin alphabet: l, m, n and t which are pronounced as "el" – "em" – "en" – "te" (in Latin it is "elementum"), probably, producing the word "element". In this way the scientists wanted to emphasize that as words are composed of letters, different compounds can be represented as constituted by elements. Such interpretation is as simple as it is unexpected. There are other explanations as well but we shall not dwell on them.

HOW "AN ELEMENT" BECAME "A CHEMICAL ELEMENT"

Before the modern model of the atom evolved, the concept of an element had been purely speculative. One of the definitions of an element belongs to Aristotle, one of the greatest philosophers of antiquity, who wrote: "Elements are simple substances of which the universe is composed and one of which cannot be separated into the other." Aristotle held that there is one primary matter and four fundamental qualities: heat and coldness, dryness and wetness. Their combinations are material elements: fire, water, air and earth. According to Aristotle, all bodies are composed of these elements. Aristotle's teaching was the theoretical foundation of alchemy and various natural philosophy schools for many centuries to come.

Only in the 16th century Paracelsus, a famous physician and scientist, brought the elements "closer to the earth". He suggested that all substances consist of three sources, mercury, salt and sulphur, which are the carriers of three qualities: volatility, solidity and inflammability.

Hints for a proper understanding of the nature of elements can be found in the teaching of Robert Boyle, an outstanding 17th century English Chemist. In his book, the skeptical chemist Boyle criticized the view of elements as carriers of certain qualities. Elements, according to Boyle, must be material in their nature and constitute solid bodies. Boyle also spoke against the belief that the number of elements is limited, thus opening up possibilities for the discovery of new elements. Nevertheless, it was still a long way to a clear understanding of what a chemical element is and therefore, scientists could not properly explain the discoveries of new elements.

Antoine Lavoisier's views were a considerable step forward in this field. He clearly stated his conception of simple bodies: he believed that all substances which scientists had failed to decompose in any way were elements and he divided all simple substances into four groups.

The first group comprised oxygen, nitrogen, hydrogen, as well as light and "thermogen" (which was, of course, a mistake). A. Lavoisier considered these simple substances to be real elements. Into the second group Lavoisier included sulphur, phosphorus, coal, a radical of muriatic acid (later called chlorine), a radical of hydrofluoric acid (fluorine), and a radical of boric acid (boron). According to Lavoisier, they all were simple non-metallic substances capable of being oxidized and of producing acids. The third group comprised simple metal substances: antimony, silver, arsenic, bismuth, cobalt, copper, tin, iron, manganese, mercury, molybdenum nickel, gold, platinum, lead, tungsten, and zinc. They also could be oxidized and form acids. And, at last, the fourth group included salt-forming compounds ("earths"), which, however, were known to be complex: lime (calcium oxide), magnesia (magnesium oxide), baryta (barium oxide), alumina (aluminium oxide) and silica (silicon oxide). In 1789, the fact that these substances are oxides of unknown elements was only a conjecture. This classification and comments were still greatly confused and unclear, but they served as a program for further research into the nature of elements.

Lavoisier did not distinguish the concepts of "an element" and "a simple body". They were clearly stated only in the 19th century owing to the development of the atomic and molecular theory and to the work of D.I. Mendeleev.

WAS THERE ANY ORDER IN THE DISCOVERIES OF ELEMENTS?

It would seem more logical to put this question towards the end of the book when the reader is already acquainted with the history of the discovery of each element. All discussions should be supported by facts and we shall do so in due time. Here we shall give only the general picture, "a bird's-eye view" of the problem so to speak.

Open pages of the book where a chronological table of the discoveries of the elements is given. Which of them were discovered in the first place? For about ten of the elements the column "Date of discovery" contains, instead of an exact date, the words "known in antiquity". The concept of antiquity is rather loose and the words mean only that these elements were known long before our time. Of course, we do not know who discovered them. Archaeologists, whose science is very far removed from chemistry, give more or less reliable information on the time when an element was used by man for the first time in antiquity (without, of course, being perceived as an element). Here is the list of elements known in antiquity: iron, carbon, gold, silver, mercury, tin, copper, lead, sulphur. Even a beginner in chemistry understands that these elements differ broadly in their properties. Why then do they occupy the first place in the list of the discoveries of elements? Is it because they are the most abundant elements on Earth?

As regards abundance, only iron and carbon are among ten of the most abundant elements. Sulphur is also fairly abundant. The remaining are rare on Earth.

Topmost in the list of the most abundant elements are oxygen, silicon, and aluminium. Man breathed oxygen unaware that it is a chemical element up to the end of the 18th century. Silicon is the earth's main material but it was discovered only in the 19th century just as aluminium although clay (alumina) had been used for ages.

All this shows that abundance of chemical elements is by no means related to the date of their discovery. Hence, the statement "the more, the earlier" is erroneous. But why were these elements known from time immemorial?

In spite of the difference in their properties, these elements have something in common. Most of them occur in nature not in the form of chemical compounds but as simple substances. For instance, even at present we come across reports of finds of gold nuggets. To find them, no chemical work is required. It is enough just to look for them. Silver and sulphur occur on Earth in a free state (but mainly as constituents of minerals); copper and mercury are encountered in a free state much less frequently. This is why these elements were among the first ones to be discovered by man. A special place is held by carbon; perhaps, it was actually the very first element which announced its existence as ashes of the first camp-fire. Iron gave its name to a whole epoch in the history of mankind – the Iron Age. Many scientists believe that our forebears first began to use iron in a free state, namely the meteorite iron. And only later did the primitive metallurgists learn to smelt iron from iron ore. Tin and lead were smelted from minerals. Extraction of these metals from compounds (the modern term is "the reduction processes") is relatively simple and could be done by people who knew next to nothing about chemical procedures.

In various regions of the globe, people began to use this or that element at different times. And, therefore, the most exact discovery date can usually be found from the first mention of an element's use. Here the term "discovery" is arbitrary and has almost nothing in common with its meaning in later time when human knowledge attained a higher level.

The age of discovery of chemical elements began only in the second part of the 18th century. Preceding millennia had seen the discovery of only five new elements: arsenic, antimony, bismuth, phosphorus and zinc. They were discovered by chance by alchemists who in vain were looking for the philosophers' stone. The peculiar properties of these elements were of great help in their manipulations (such as, for instance, amazing luminescence of phosphorus in the dark and unusual features of arsenic compounds).

The discovery of new chemical elements became a routine matter and not a stroke of good luck only after two main conditions had been met. First of all, chemistry had begun to take shape as an independent science, its experimental methods had become satisfactory and scientists had learned how to determine the composition of minerals, those treasure troves of chemical elements. Secondly, most scientists came at last to a consensus on the conception of a chemical element. It was the beginning of a great analytical period in the development of chemistry in the course of which a large part of naturally-occurring elements were discovered.

Particularly interesting is the story of the discovery of hydrogen and elementary atmospheric gases, nitrogen and oxygen. It became possible owing to the progress in pneumatic chemistry. For a long time, the study of gases was the privilege of physicists and for a long time discoverers of new gases believed that they were only varieties of air. The realization that these varieties are chemical elements was slow in coming. It was, first of all, necessary to review cardinally the old theoretical conceptions and to reject the so-called theory of phlogiston, which was believed to be the primary matter of combustion. We shall come back to the phlogiston theory later. These efforts of scientists brought due rewards: the discovery of nitrogen, hydrogen and oxygen played a gigantic role in advancing the most important concepts of modern chemistry, its theoretical foundations and experimental methods.

Thus, it does not seem paradoxical any more that oxygen (the most abundant element constituting almost one half of the earth's crust by mass) was discovered so late. Chemistry had to stand firmly on its feet to be able to identify oxygen as a new simple substance. Adequate methods of investigation were required for this purpose.

Various analytical methods, constantly perfected, were the key factors which led, step by step, to the discovery of new chemical elements. But chemical analysis by itself was not enough to fill all the boxes in the periodic table. The scientists divined the existence of many new elements not because they discovered them, figuratively speaking, on the bottom of a test tube. These elements made their existence in nature known in another way (especially those of them whose abundance is very low).

Billions of years were required for the formation of the earth's crust with its minerals and ores, a process bearing witness to many whims of nature which, to be more exact, reflect the laws of geochemistry. Some elements were less fortunate: they did not succeed in forming their minerals, that is, those in which they would be the principal or, at least, a noticeable component. They exist only as admixtures to all sorts of minerals consisting of other elements. They seem to be widely dispersed in the earth's crust and are called "trace" elements. Only in the rarest cases do they form their minerals and if scientists were lucky to come across them, the new element immediately became the target of chemical analysis. As we shall see later, this was the case of germanium extracted from argyrodite, a uniquely rare mineral.

The other trace elements have quite a different history. Cesium, rubidium, indium, thallium and gallium are classic examples of new chemical elements which were identified at first without the help of chemistry. They announced themselves with the aid of a peculiar visiting card - their spectrum. It was spectral analysis, a new research method that contributed to their discovery. If a grain of a substance is introduced into the flame of a gas burner and the light passes through a prism, the refracted light contains several differently arranged spectral lines of various colours. Studying the spectra of known elements, scientists concluded that each element has its spectral picture. Spectral analysis at once showed itself as a powerful research tool. If the spectrum of a compound contained unknown lines, it was logical to assume that this compound contained a new element. Cesium, rubidium, indium thallium and gallium was discovered exactly in this manner. However, in such cases it took courage for scientists to announce the existence of new elements since they had not a grain of them in their hands and did not know their properties.

Such unusual chemical elements as helium, neon, argon, krypton and xenon were discovered by their spectra. They were termed noble or inert gases. Their content in the atmosphere is extremely low. For a long time, these gases were considered to be quite incapable of chemical reactions and some even believed that the name of "a chemical element" was inapplicable to them. Inert elements were discovered without the aid of chemistry but their extraction from the atmosphere and separation from one another became possible only after the development of methods of gas liquefaction at low temperatures.

Naturally, the history of the discovery of chemical elements was to an extent affected by the abundance factor: the elements less abundant in nature were discovered later. The history of natural radioactive elements gives a fine illustration of this idea. They were discovered at the end of the 19th and beginning of the 20th century. And if it had not been for a very important event, they would have remained unknown to mankind for a long time since neither chemical nor spectral methods of analysis could detect the negligible concentrations of these elements. The event was the discovery of a new physical phenomenon called radioactivity. Some substances can spontaneously and continuously emit radiation. At first, it was established that this property is peculiar not to these substances in general and even not to the constituent chemical compounds but to specific chemical elements, uranium and thorium, placed at the very bottom of the periodic table. In the studies of radioactive substances, it was noticed that sometimes their radiation is much stronger than that typical of uranium and thorium atoms. It was suggested that this radiation was due to unknown radioactive elements. The suggestion was confirmed by the discovery of polonium and radium. This led to another research method – the radiometric method – which, in the long run, led to the discovery of other natural radioactive elements. In this example, radioactivity served as an identification mark. The radiometric method is incomparably more sensitive than other methods of detection of elements.

After the late twenties of our century, there were no more discoveries of the elements existing in nature. But this was not the end of the history of discoveries of new elements. However, the word "discovery" acquired a new meaning. It now referred to elements not existing on Earth but prepared artificially by means of nuclear reactions. It was a problem of extreme scientific and technical complexity which was tackled by scientists of many countries. All artificial or synthesized elements are radioactive and therefore, the radiometric method has played the most important role. Here the decisive word was said by physicists. But chemists were confronted with a very difficult problem. Even in our time many synthesized elements can be obtained in the amounts of just a few atoms. When these atoms are highly radioactive, their lifetime is only a fraction of a second. Therefore, chemists must show miracles of inventiveness to study their properties.

This, in a nutshell, is the centuries-long process of discovery of the chemical elements, whose symbols now appear in the Mendeleev's periodic table. We shall consider this process in detail. Let us now have a closer look at the principal characters of this narrative – one at a time.

But, first, a few words about the structure of the book. It consists of two parts. The first part deals with the natural elements, the second part – with the synthesized ones. It is obvious that the first part must begin with the description of the elements known in antiquity (Chapter 1); then we shall dwell on the elements discovered in the Middle Ages (Chapter 2). The term "discovery" cannot properly be applied to the elements described in these chapters. It acquired the present-day meaning only after the concept of "a chemical element" was made more precise. This was facilitated by the progress in pneumatic chemistry and by a gradual refutation of the phlogistic theory and accompanied by the discovery of oxygen, nitrogen and hydrogen as well as the understanding of their elementary nature (Chapter 3).

A considerable number of new chemical elements were discovered in the second half of the 18th century and the first half of the 19th century with the aid of chemical analysis (Chapter 4); the electrochemical method played a certain role in the separation of some alkali and alkaline - earth metals (Chapter 5). At the turn of the fifties of the last century, the spectroscopic method was developed, with the help of which it became possible to identify several new elements on Earth (Chapter 6).

Of special interest is the discovery of rare earths, noble (or inert) gases and finally, the elements predicted by D.I. Mendeleev on the basis of the periodic system. Although these elements were discovered using chemical analysis and spectroscopic method, the histories of the above groups of elements are in many respects highly individual and separate chapters have been devoted to their presentation (chapters 7, 8 and 9). No less peculiar is the history of the two stable elements which proved to be the last to be discovered on Earth - hafnium and rhenium (chapter 10). The first part of the book ends with the history of radioactive elements (chapter 11), which introduces the reader to the world of radioactivity, the world of unstable elements and isotopes, the most of which were obtained artificially by means of nuclear reactions.

The second part of the book comprising two chapters (chapters 12 and 13) is devoted to synthesized elements. In chapter 12, the reader will be introduced to the synthesis of new elements within the previous boundaries of the periodic system – from hydrogen to uranium (technetium, promethium, astatine, francium). Chapter 13 covers the history of transuranium elements and prospects of nuclear synthesis.

Part - I

ELEMENTS DISCOVERED IN NATURE

Most chemical elements known at present have been discovered in nature (in various ores and minerals, the earth's atmosphere, etc.) and one can say with confidence that there are no more undiscovered elements in nature, including both stable elements and those referred to as naturally radioactive ones. They can be called elements "discovered by means of analysis". They exist independently of man, his knowledge and methods of investigation. They existed at the earliest stages of evolution of the solar system when the Earth was being formed as a planet.

How these elements were discovered is the subject of the first part of our book.

More than 90 percent of elements occurring in nature are stable, i.e. not radioactive. They occupy boxes from 1 to 83 in the periodic table, i.e. from hydrogen to bismuth. There are two gaps in this sequence corresponding to the elements with $Z = 43$ (technetium) and $Z = 61$ (promethium). The strange properties of atomic nuclei have made all the isotopes of these elements radioactive with relatively short lifetimes; therefore, technetium and promethium have not been preserved in nature but decayed and transformed into the neighbouring stable elements.

The number of naturally radioactive elements on Earth is considerably smaller than that of stable ones. In the periodic table, they begin with polonium (z = 84) and end with uranium (z = 92). Among them, only thorium and uranium have very long half-lives; therefore, they have survived on Earth since the time of their formation and their amounts are rather noticeable. That is why uranium and thorium have been discovered as new chemical elements long before scientists succeeded in observing radioactivity. The amounts of other naturally radioactive elements (polonium, radon, radium, actinium and protactinium) are much smaller.

1

Elements known in Antiquity

Antiquity is, of course, a loose concept and, therefore, this heading under which we discuss several chemical elements is, to a great extent, arbitrary, though it has been widely used in history. This chapter deals with elements (mainly, metals), the use of which is either mentioned in various written sources of the distant past or can be established from the archaeological data.

The use of the term "discovery" in this case is quite arbitrary. Historically speaking, the principal characters of this chapter were recognized as independent chemical elements relatively recently. A description of the early history of the elements of antiquity will of necessity have to pass over in silence the dates and the authors of the discoveries. Therefore, the manner of presentation of material in this chapter is rather unusual. It will be a short report on these elements and their application in the distant past.

The chapter is devoted to seven metals of antiquity: gold, silver, copper, iron, tin, lead and mercury – the "magnificent seven" of metals that played a tremendous role both in the development of civilization and in various schools of natural philosophy. We shall tell you about sulphur, which was widely used long before our time and about carbon. It may well be that carbon is the oldest chemical element known to mankind. Therefore, we shall begin the history of chemical elements with carbon.

Sometimes zinc, platinum, antimony and bismuth are also said to be known in antiquity but there is no definite proof of that.

CARBON

Carbon (Latin carbo, carbonaris "coal") non-metallic element, symbol C, atomic number 6, relative atomic mass 12.011. It occurs on its own as diamond, graphite and as fullerenes, as compounds in carbonaceous rocks such as chalk and limestone, as carbon dioxide in the atmosphere, as hydrocarbons in petroleum, coal and natural gas and as a constituent of all organic substances.

In its amorphous form, it is familiar as coal, charcoal and soot. Of the inorganic carbon compounds, the chief ones are carbon dioxide, a colourless gas formed when carbon is burned in an adequate supply of air and carbon monoxide (CO), formed when carbon is oxidized in a limited supply of air. Carbon disulphide (CS_2) is a dense liquid with a sweetish odour.

Another group of compounds is the carbon halides, including carbon tetrachloride (tetrachloromethane, CCl_4).

When added to steel, carbon forms a wide range of alloys with useful properties. In pure form, it is used as a moderator in nuclear reactors; as colloidal graphite it is a good lubricant and when deposited on a surface in a vacuum, obviates photoelectric and secondary emission of electrons. Carbon is used as a fuel in the form of coal or coke. The radioactive isotope carbon-14 (half-life 5,730 years) is used as a tracer in biological research.

The element has the following characteristic reactions.

With air or oxygen: It burns on heating to form carbon dioxide in excess air, or carbon monoxide in a limited supply of air.

$$C + O_2 \rightarrow CO_2$$
$$\Delta H = -394 \; KJ \; mol^{-1}$$
$$2C + O_2 \rightarrow 2CO$$

With metal oxides: It reduces many metal oxides at high temperatures.

$$Fe_2O_3 + 3C \rightarrow 2Fe + 3CO$$

With steam: It forms water gas (a cheap, useful, industrial fuel) when steam is passed over white-hot coke.

$$C + H_2O \rightarrow Co + H_2$$

With concentrated acids: with hot, concentrated sulphuric or nitric acid, it forms carbon dioxide.

The exact date of the discovery of carbon cannot be ascertained. However, it is not difficult to find out when carbon was identified as a simple substance. Let us direct our attention to the "Table of Simple Bodies" compiled by A. Lavoisier and published in 1789. Carbon appears as a simple substance in it. However, the time that carbon needed to occupy its place in the table is measured not by years and even not by centuries but by millennia. Man had met carbon even before he could make fire - in the form of woods burnt by lightning. After man had learned how to start a fire, carbon became his constant "Companion".

Carbon played an important role in the progress of the phlogistic theory. According to this theory, carbon was not a simple substance but pure phlogiston. By studying the combustion of coal and other compounds, A. Lavoisier was the first to show that carbon is a simple substance. Here we are going to digress a little from the story about how carbon found its identity.

In nature carbon occurs in two allotropic modifications – diamond and graphite, both known to man for a long time. The fact that diamond burns without a residue at very high temperatures was also known long ago. Nevertheless, diamond and graphite were believed to be two quite different substances. The discovery of carbon dioxide was an event which helped to establish that diamond and graphite are modifications of the same substance. After experimenting with the burning of diamond and charcoal, A. Lavoisier established that upon combustion both substances yield carbon dioxide. This prompted the conclusion that diamond and coal have the same origin. The name "Carboneum" (carbon) appeared for the first time in the book "Methods of chemical Nomenclature" (A. Lavoisier, L. Guyton de Morvean, C. Berthollet and A. Fourcroy) in 1787.

A parallel can be drawn between the element itself, known from time immemorial and its Latin name whose root originates from Sanskrit, one of the oldest known languages. In Sanskrit, "Cra" means "to boil". The name "carbon" was suggested in 1824.

In 1797, S. Tennant discovered that combustion of equal amounts of diamond and graphite liberate equal amounts of carbon dioxide. In 1799 L. Guyton de Morveau confirmed that carbon is the only constituent of diamond, graphite and coke. Twenty years later, he succeeded in transforming diamond into graphite and then into carbon dioxide by careful heating. But the reverse transformation of graphite into diamond was beyond the power of the science of the 18th and 19th centuries. It was only in 1955 that a group of English scientists obtained artificial diamonds for the first time in the world's history. Synthesis was performed at 3000^0 C under a pressure exceeding 10^0 Pa.

Soon after the synthesis of diamond, Soviet scientists prepared a new substance, carbine, which, has since been proved, is the new, third allotropic modification of carbon. The carbon atoms in it comprise long chains. This substance resembles soot.

The study of carbon and its compounds laid the foundation of a vast field of chemistry – organic chemistry.

SULPHUR

Sulphur brittle, pale-yellow, non-metallic element, symbol S, atomic number 16, relative atomic mass 32.064. It occurs in three allotropic forms: two crystalline (called rhombic and monoclinic, following the arrangements of the atoms within the crystals) and one amorphous. It burns in air with a blue flame and a stifling odour. Insoluble in water but soluble in carbon disulphide, it is a good electrical insulator. Sulphur is widely used in the manufacture of sulphuric acid (used to treat phosphate rock to make fertilizers) and in making paper, matches, gunpowder and fireworks, in vulcanizing rubber and in medicines and insecticides.

It is found abundantly in nature in volcanic regions combined with both metals and non-metals and also in its elemental form as a crystalline solid. It is a constituent of proteins and has been known since ancient times.

Sulphur has been known to man for a very long time. Even in times of Homer, ancient Greeks used the specific properties of sulphur dioxide liberated in the burning of sulphur for disinfection of homes. Deposits of native sulphur have also been known from ancient times. Thus, Pliny the Elder described the deposits of sulphur in Italy and Sicily. Sulphur was used for making dyes and treating fabrics. Like carbon, from the earliest times sulphur was used in pyrotechnics. The composition known by the name of "Greek fire" and invented, apparently, in the 5th century A.D. in Byzantium was a mixture of finely ground sulphur (one part), coal (two parts) and saltpetre (six parts). It is interesting to note that this composition differs only slightly from that of black (smoky) gunpowder.

The fact that sulphur is a good combustible material and combines readily with a great number of metals is responsible for its "privileged" position among other substances in the middle ages. Alchemists considered sulphur as the element of combustibility and a constituent of all metals. Very unusual properties were often attributed to sulphur, although some alchemists described its real properties rather accurately.

The elemental nature of sulphur was established by A. Lavoisier. However, in spite of the fact that by the beginning of the 19th century, sulphur had already been recognized as an independent element, experiments had to be carried out to elucidate the exact composition of native sulphur. In 1808, H. Davy suggested that sulphur in its usual state is a combination of small amounts of oxygen and hydrogen with a great amount of sulphur. This questioned the elemental nature of sulphur but in 1809, Gay Lussac proved it beyond any doubt. In 1810 H. Davy pointed out that the presence of oxygen in sulphur may be due to sulphur oxides present in native sulphur. The oxygen content in sulphur varied depending on the deposit where the samples were taken. From the standpoint of modern chemistry, one may say that oxygen found by Davy in sulphur was not the oxygen of sulphur oxides but that of oxysulphides of various metals, which are always present in sulphur.

The origin of the Latin word "sulphur" is unclear.

GOLD

Gold - heavy, precious, yellow, metallic element; symbol Au, atomic number 79, relative atomic mass 197.0. It is unaffected by oxygen and is highly resistant to acids. For manufacture, gold is alloyed with another strengthening metal (such as copper or silver), its purity being measured in carats on a scale of 24.

In 1990 the three leading gold-producing countries were South Africa, 605.4 tonnes; USA, 295 tonnes; and Russia, 260 tonnes. In 1989, gold deposits were found in Greenland with an estimated yield of 12 tonnes per year.

Gold occurs naturally in veins, but following erosion it can be transported and re-deposited. It has long been valued for its durability, malleability and ductility and its uses include dentistry, jewelry and electronic devices.

Karl Marx wrote: "Gold is in fact the first metal that man has discovered."

This is really so. Gold articles were found in excavations together with stone tools dating from the Neolithic Age. But in those times people, evidently, used gold found by chance. Only after the emergence of classes in society, first attempts were made to mine gold. The explanation is simple. Gold was particularly suited to play the function of money due to its properties of immutability, easy divisibility and high cost.

As an ornamental material, gold began to be used from time immemorial. During excavations of pyramids of all dynasties in Egypt, archaeologists found in great numbers not only gold jewelry but also household articles.

Gold was known not only in Egypt. As early as in the 10th century B.C. it was used in China, India, and States of Mesopotamia. In Greece, gold coins circulated as far back as in the 8th-7th centuries B.C. In Armenia, gold coins appeared in the 1st century B.C. Thus, gold was known to the people of ancient states in Europe and Asia. The oldest gold mines were found in India and Nubia (North-East Africa).

The processes of gold purification known in antiquity did not yield the pure metal but usually alloys consisting of gold and silver which were named azem. A natural gold - silver alloy - electrum - was also known.

No other metal has played so sinister a role as gold in the history of mankind. Wars were waged, nations and states were annihilated, and monstrous crimes were committed for the sake of gold. But possession of gold did not bring peace to man. On the contrary, sorrow and fear of losing this treasure filled his soul.

The alchemic period between the 4th and the 16th centuries was a gloomy one in the history of the search for gold. The efforts of alchemists were directed towards the search for the "philosophers' stone" which, they held, possessed the property of transforming base metals into gold. Alchemy did not start from scratch but had important precursors. Egypt's fast rise was due to the fact that Egyptians possessed the secret of gold extraction. It was also known that iron articles that remained in copper mines for a long time became coated with copper. Iron was believed to transform into copper. If it was so, why could not other metals be transformed into gold? Native lead sulphide almost always contains an admixture of silver, which could sometimes be extracted. Could not silver be formed on lead? And, finally, progress in alchemy was facilitated by the idea about the unity of matter according to which all substances consist of the same components in different ratios.

All the attempts to find the "philosophers' stone" turned out to be unsuccessful (as one should have expected), although many alchemists gave their lives for the idea. All reports about the discoveries of methods of preparing gold from other metals were pure charlatanism.

Alchemy was still flourishing in Europe when the first Spanish conquistadors set out for South and Central America. In the land of Incas, they were amazed by the tremendous amounts of gold. For Incas gold was a sacred metal, the Sun God's metal and colossal amounts of gold had accumulated in the temples. When the Spaniards took Atahualpa, the Great Inca, prisoner, they promised him freedom for a fantastic ransom of almost 50 m³ of gold. But Francisco Pizarro thought it dangerous to free the Great Inca and without waiting for the ransom, the Spaniards executed Atahualpa. When the Incas learned about the death of their leader, the caravan consisting of 1100 llamas carrying gold had already been on its way. Incas hid the gold in the mountains of Azangaro ("the remotest place"). But they could not hide all their treasures. Spaniards captured and looted Cuzco, one of Peru's richest cities. They melted the priceless creations of ancient craftsmen into gold ingots and sent them to Spain.

In Russia, mining of gold began in 1600 but it was not until the 19th century that large-scale extraction of this metal started.

The Latin name for gold, aurum, originates from the word Aurora (dawn).

SILVER

Silver - white, lustrous, extremely malleable and ductile, metallic element, symbol Ag (from Latin argentum), atomic number 47, relative atomic mass 107.868. It occurs in nature in ores and as a free metal; the chief ores are sulphides, from which the metal is extracted by smelting with lead. It is one of the best metallic conductors of both heat and electricity; its most useful compounds are chloride and bromide, which darken on exposure to light and are the basis of photographic emulsions.

Silver is used ornamentally, for jewelry and tableware, for coinage, in electroplating, electrical contacts and dentistry and as a solder. It has been mined since prehistory; its name is an ancient non-Indo-European one, silubur, borrowed by the Germanic branch as silber.

Silver is a more active metal than gold but, although its abundance in the earth's crust is about fifteen times that of gold, it occurs much less frequently in a native state. It is not surprising that in antiquity silver was valued higher than gold. In ancient Egypt, for instance, the ratio between the costof these metals was 2.5:1. Gold was used mainly for coins and jewelry; silver had other uses: for example, for making water vessels.

In the 4th century B.C. the army of Alexander the Great conquered Persia and Phoenicia and invaded India. Here the Greek army was struck by an outbreak of a mysterious gastrointestinal disease and the men demanded to be sent home. Interestingly, the Greek military commanders fell victim to the disease far less frequently than their men, although they shared all the burdens of camp life with the soldiers. More than two thousand years had passed before scientists found an explanation of it. The soldiers drank from tin cups and their superiors from silver ones. It was proved that silver dissolves in water forming a colloid solution that kills pathogenic bacteria. And although the solubility of silver in water is low, it is quite enough for disinfection.

Silver mines have been known from time immemorial. The largest deposits of silver were in Greece, Spain and Germany. After the discovery of America, silver deposits were also found in Peru and Mexico. Lead minerals are often observed as constituents in silver ores. An old process of extracting silver from such ores is described as follows. Silver ore was ground, washed with water and dried. Then it was fused together with flux and the alloy thus obtained was heated with Charcoal. The resulting alloy of silver and lead was calcinated. On heating in air, silver is practically unoxidized whereas lead transforms into oxide almost completely. The melting point of lead oxide is 896°C and that of silver, 960°C. Thus, practically pure silver was obtained. At present, more perfect processes of purifying silver are used.

Silver like gold was used in coins but the cost of silver compared to that of gold was gradually decreasing. In 1874, the cost of one pound of gold was equal to that of 15.5 pounds of silver but after the discovery of silver deposits in Australia this ratio fell to 1:46. In England bimetallism, i.e. the use of gold and silver jointly as a monetary standard, was discontinued in 1816.

The name "Silver" seems to stem from the Assyrian "Serpu" or Gothic "Silbur". The Latin argentum originates most likely from the Sanskrit arganta, which means "light, white".

COPPER

Copper - orange pink, very malleable and ductile, metallic element, symbol Cu (from Latin cuprum), atomic number 29, relative atomic mass 63.546. It is used for its durability, pliability, high thermal and electrical conductivity and resistance to corrosion.

It was the first metal used systematically for tools by humans; when mined and worked into utensils it formed the technological basis for the Copper Age in prehistory. When alloyed with tin it forms bronze, which strengthens the Copper, allowing it to hold a sharp edge; the systematic production and use of this was the basis for the prehistoric Bronze Age. Brass, another hard copper alloy, includes zinc. The element's name comes from the Greek for Cyprus (Kyprios), where copper was mined.

According to the French chemist M. Berthelot, mankind came to know copper more than five thousand years ago. Other scientists believe that this acquaintance is much older. Copper and its alloy with tin (bronze) had for a long time been the most widely used metals. These two materials marked a whole epoch in the history of mankind, the Bronze Age. Why did copper play such an important part? Copper is fairly abundant in nature and can readily be worked. At first people used only native copper but later rising demand led to the processing of copper ores. It is comparatively simple to smelt the metal from ores with high copper content. As early as the third millennium B.C. copper was widely used for manufacturing various tools. The Egyptian Pyramid of Cheops was built with gigantic stone blocks each of which was hewn with copper tools.

Among the copper mines of antiquity, the particularly famous ones were those on the island of Cyprus to which, as has been suggested, copper owes its name (cuprum in Latin).

Only when man had learned to produce bronze, stone tools were completely replaced with bronze ones. Most likely bronze was first obtained by chance. This is evidenced by the archaeological finds on the island of Crete dating back to about 3500 B.C. which revealed not only copper but bronze articles as well, at first bronze was rather expensive and was used mainly for jewelry and luxury articles. In ancient Egypt mirrors were made from bronze. Bronze, like copper, proved to be an excellent material for relict makers and sculptors. As early as the 5th century B.C., man learned to cast bronze statues. Particular progress in bronze sculpture was made in ancient Greece beginning with the Mycenaean period. At our times copper and bronze still retain this role.

Besides bronze, another wonderful copper alloy, brass, has been known for a long time. It was prepared by fusing copper with zinc ore. Ancient Egyptians, Indians, Assyrians, Romans and Greeks knew copper, bronze and brass. Both copper and bronze were used for making weapons. In excavations dated back to the 8th-6th centuries B.C. in Altai, Siberia and Trans–Caucasus archaeologists found knives, arrow–heads, shields and helmets made from bronze and copper. In ancient Greece and Rome, copper and bronze were also used for making shields and helmets. Copper found other uses in firearms when they had been invented.

IRON

Iron - hard, malleable and ductile, silver–grey, metallic element, symbol Fe (from Latin ferrum), atomic number 26, relative atomic mass 55.847. It is the fourth most abundant element (the second most abundant metal, after aluminium) in the Earth's crust. Iron occurs in concentrated deposits as the ores hematite (Fe_2O_3), spathic ore ($FeCO_3$) and magnetite (Fe_3O_4). It sometimes occurs as a free metal, occasionally as fragments of iron or iron-nickel meteorites.

Iron is the most common and most useful of all metals; it is strongly magnetic and is the basis for steel, an alloy with carbon and other elements. In electrical equipment, it is used in all permanent magnets and electromagnets and forms the cores of transformers and magnetic amplifiers. In the human body, iron is an essential component of haemoglobin, the molecule in red blood cells that transports oxygen to all parts of the body. A deficiency in the diet causes a form of anaemia.

Iron is the second most abundant metal in nature after aluminium. But native iron is extremely rare. Probably, the first iron used by our forefathers was of a meteoritic origin. Iron oxidizes readily in the presence of water and air and is found mainly in the form of oxides. Oxidation of iron is responsible for the fact that extant articles made of iron in antiquity are extremely rare. Man discovered iron about five thousand years ago. At first iron was very expensive and was valued much higher than gold; very often iron jewelry was set in gold.

People of all continents became aware of gold, silver and copper approximately at the same time; but in the case of iron the situation is different. Thus, in Egypt and Mesopotamia the process of extracting iron from ores was discovered two thousand years B.C.; in Trans-Caucasus, Asia Minor and ancient Greece at the end of the second millennium; in India in the middle of the second millennium and in China much later, only in the middle of the first millennium B.C. In the countries of the New World, Iron Age began only with the arrival of Europeans, i.e. in the second millennium A.D.; some African tribes began to use iron skipping the Bronze Age period in development. This is due to the difference in natural conditions. In countries where natural resources of copper and tin were small, a demand arose for replacing these metals. America had one of the largest deposits of native copper and therefore, it was not necessary to search for new metals. Gradually, production of iron grew and iron began to pass from the category of precious metals into that of ordinary ones. By the beginning of the Christian era, iron was already widely used.

Among all metals and alloys known by that time, iron was the hardest one. Therefore, as soon as iron grew relatively cheap, various tools and weapons were manufactured from it. At the beginning of the first millennium A.D. production of iron in Europe and Asia had made considerable progress; particularly great successes in smelting and processing iron had been achieved by Indian metallurgists.

It is interesting to have a look at the development of iron production methods. At first man used only meteoritic iron, which was very rare and therefore expensive. Then people learnt how to produce iron by intensively heating its ores with coal on windy sites. Iron thus obtained was spongy, of low grade and with large inclusions of slag. An important step in iron production was made with the invention of a furnace open at the top and lined with a refractory material inside. Excavations of ancient towns in Syria indicate that iron of a rather good quality was produced in this way. Later, people noted that cast iron, which had been considered to be a waste product, could be transformed into iron, the process requiring much less coal and yielding high quality iron.

By the end of the 15th century, first smelting furnaces appeared producing exclusively cast iron. Iron and steel smelting processes were rapidly improving. In 1855, there appeared the converter process of steel-making which is still used. The martin process developed in 1865 yields steel almost free of slag.

A chemical symbol Fe originates from the Latin ferrum, which means "iron".

LEAD

Lead - heavy, soft, malleable, grey, metallic element, symbol Pb (from Latin plumbum), atomic number 82, relative atomic mass 207.19. Usually found as an ore (most often in galena), it occasionally occurs as a free metal (native metal) and is the final stable product of the decay of uranium. Lead is the softest and weakest of the commonly used metals, with a low melting point; it is a poor conductor of electricity and resists acid corrosion.

As a cumulative poison, lead enters the body from lead water pipes, lead-based paints and leaded petrol. (In humans, exposure to lead shortly after birth is associated with impaired mental health between the ages of two and four). The metal is an effective shield against radiation and is used in batteries, glass, ceramics and alloys such as pewter and solder.

Lead is very rarely encountered in a native state but is smelted fairly easily from ores. Lead became known to Egyptians simultaneously with iron and silver and was produced as early as the second millennium B.C. in India and China. In Europe, production of lead began somewhat later, although in the 6th-Century B.C. records we find mention of lead which was brought to the tyre trade fair. Lead was produced in great amounts during the reign of Hammurabi in Babylon. For a long time, lead was confused with tin. Tin was named "plumbum album" and lead – "plumbum nigrum". Only in the Middle Ages were they recognized as different metals.

Greeks and Phoenicians started many lead mines in Spain which later were taken over by Romans. In ancient Rome lead was widely used for making crockery, styluses and pipes for the famous Roman water-main. Lead was also used for manufacturing white lead. The island of Rhodes was the biggest exporter of white lead. The process of its preparation is still used as follows: lead pieces are immersed into vinegar and the salt thus obtained is boiled with water for a long time. But red lead was first obtained unexpectedly. When a fire broke out in the Greek port of Piraeus, barrels with lead were enveloped in flames. After the fire had been extinguished, red substance was found in the charred barrels - it was red lead.

Although in Russia, lead has been known for a long time, up to the 18th Century the process of lead production was very primitive. After the invention of firearms, lead was used for making bullets and the military importance of lead is still great. But in addition to its "military" uses, lead has many peaceful ones; for instance, typographical types are made of its alloy with antimony. Lead is also used for protection against radiation in experiments.

Greeks named lead molibdos; its chemical symbol pb originates from latin plumbum.

TIN

Tin - soft, silver-white, malleable and somewhat ductile, metallic element, symbol Sn (from Latin Stannum), atomic number 50, relative atomic mass 118.69. Tin exhibits allotropy, having three forms: the familiar lustrous metallic form above 55.8°F/13.2°C; a brittle form above 321.8°F/161°C; and a grey powder form below 55.8°F/132°C (commonly called tin pest or tin disease). The metal is quite soft (slightly harder than lead) and can be rolled, pressed, or hammered into extremely thin sheets; it has a low melting point. In nature, it occurs rarely as a free metal. It resists corrosion and is therefore used for coating and plating other metals.

Tin and copper smelted together form the oldest desired alloy, bronze. Since the Bronze Age (3500 BC) that alloy has been the basis of both useful and decorative materials. The mines of Cornwall were the principal western source from then until the 19th Century, when rich deposits were found in South America, Africa and S.E. Asia. Tin is also alloyed with metals other than copper to make solder and pewter. It was recognized as an element by Antoine Lavoisier, but the name is very old and comes from the Germanic form zinn.

Tin typically occurs in nature in the form of the mineral cassiterite. It is believed that man discovered tin about 6-6.5 thousand years ago, i.e. in the same period as copper. Tin was widely known in the Mediterranean countries, Persia and India. Egyptians imported tin for the production of bronze from Persia. In his book Ancient Egyptian Materials and Their Production, A. Lukas writes that although in Egypt tin ores were not known, the oldest known tin articles were found in burial sites of the 18th dynasty (1580-1350 B.C.) (in particular, a ring and a vessel). Tin was known not only in the countries of the Mediterranean. Julius Caesar mentioned production of tin in central regions of Britain. Cortez, when he arrived in South America in 1519, found that tin coins were widely circulating in Mexico. However, the time of discovery of tin in America is not known.

In antiquity, tin was used not only as a component of bronze but also for making crockery and jewelry. Pliny the Elder and

Dioskorides mention tinning of copper plates to protect them from corrosion.

Up to the 13th Century, England was the only country in Europe where tin was produced. Tin was fairly expensive. In mid-16th Century, its cost was equal to that of silver and it was used for manufacturing luxury goods. Then, as its production increased, it found many applications, for instance, for making tin plate.

The Latin for tin (Stannum) stems from the Sanskrit stan which means "Solid". The chemical symbol Sn originates from the Latin name.

MERCURY

Mercury or quicksilver - heavy, silver-grey, metallic element, symbol Hg (from Latin hydrargyrum), atomic number 80, relative atomic mass 200.59. It is a dense, mobile liquid with a low melting point (–38.87°C/–37.96°F). Its chief source is the mineral cinnabar, HgS, but it sometimes occurs in nature as a free metal.

Its alloys with other metals are called amalgams (a silver-mercury amalgam is used in dentistry for filling cavities in teeth). Industrial uses include drugs and chemicals, mercury-vapour lamps, are rectifiers, power-control switches, barometers and thermometers.

Mercury is a cumulative poison that can contaminate the food chain and cause intestinal disturbance, kidney and brain damage and birth defects in humans. (The World Health organization's 'Safe' limit for mercury is 0.5 milligrams of mercury per kilogram of muscle tissue). The discharge into the sea by industry of organic mercury compounds such as dimethyl mercury is the chief cause of mercury poisoning in the latter half of the 20th century. Between 1953 and 1975, 684 people in the Japanese fishing village of Minamata were poisoned (115 fatally) by organic mercury wastes that had been dumped into the bay and had accumulated in the bodies of fish and shellfish.

The element was known to the ancient Chinese and Hindus and is found in Egyptian tombs of about 1500 BC. It was named by the alchemists after the fast-moving god, for its fluidity.

There is a science-fiction story by a Russian Scientist I.A. Efremov, the Lake of the Mountain Spirits. Anybody who visited the lake in a sunny weather died. People living in the area were sure that the lake was inhabited with evil spirits who hated all visitors. When geologists reached the lake high in the mountains, they were amazed to learn that the lake contained not only water, but also native mercury. And the "evil spirits" were nothing but mercury vapour; in hot weather they rose above the surface of small and large mercury pools surrounding the lake.

Indeed, mercury is often found in native state, sometimes in the most unexpected places. For instance, in some mountain regions of Spain mercury was found at the bottom of wells. In antiquity, mercury was known in China and India. Mercury was also found in excavations of Egyptian tombs dating from about the middle of the second millennium B.C. Most researchers believe that cinnabar was the only mercury-containing mineral known in antiquity. Theophrastos (300 B.C.) described the process of extracting mercury from cinnabar by treating it with copper and vinegar. Man discovered mercury in ancient times owing to the fact that it is comparatively easily liberated from cinnabar at a sufficiently high temperature.

The world's biggest mercury deposit is at Almaden (Spain). Exploitation of this deposit began at the time of the Roman Empire and Romans extracted 4.5 tons of mercury annually.

In antiquity, mercury had many uses. Mirrors were made with amalgamated mercury; mercury and its compounds were used as medicines. Cinnabar was mainly used as a pigment and not for producing pure mercury. Before the invention of the galvanization process, mercury had been used in gilding and silvering processes. Amalgam of the metal was applied to a metal plate and heated to a high temperature. When mercury evaporated, a thin coat of gold or silver remained on the plate. But this process was very unhealthy. Mercury played an important role in studies of gases; it was used in gas pumps and gas vessels.

Aristotle named mercury "liquid silver" and Dioskorides named it "Silver water". From this comes the Latin name of mercury - hydrargium.

2

Elements discovered in the middle ages

There are several chemical elements the history of whose discovery is not clear. We had every reason to classify the nine elements described in chapter 1 as the elements of antiquity. For the five elements - phosphorus, arsenic, antimony, bismuth and zinc - discussed in this chapter, there is evidence that people knew these elements (with the exception of phosphorus), or at least their ores and minerals, in prehistoric times, or at any rate before the Christian era. But the knowledge of them was confused and ambiguous. It became better much later, at the time of alchemistry when various chemical procedures were performed in laboratories and chemist's shops. Although their nature remained unclear, they were a basis of many useful compounds (particularly, acids and salts).

Medieval chemists discovered the elements that we shall describe in this chapter. But analytical chemistry was as yet non-existent and the bare acquaintance with these elements cannot be described as their discovery.

Thus, phosphorus, arsenic, antimony, bismuth and zinc have unusual histories. By a strange caprice of nature, P, As, Sb, and

Bi are in the main subgroup of the fifth group of the periodic table and the similarity in their properties often resulted in confusion.

As the order in which these elements were discovered is not very important, we shall begin our discussion with phosphorus.

PHOSPHORUS

Phosphorus (Greek phosphoros "bearer of light") - highly reactive, non-metallic element, symbol P, atomic number 15, relative atomic mass 30.9738. It occurs in nature as phosphates (commonly in the form of the mineral apatite) and is essential to plant and animal life. Compounds of phosphorus are used in fertilizers, various organic chemicals, for matches and fireworks and in glass and steel.

Phosphorus was first identified in 1674 by German alchemist Hennig Brand (C.1630-?), who prepared it from urine. The element has three allotropic forms: a black powder, a white-yellow, waxy solid that ignites spontaneously in air to form the poisonous gas phosphorus pentoxide and a red-brown powder that neither ignites spontaneously nor is poisonous.

Interestingly, among all elements of antiquity and Middle Ages only phosphorus has the exact (within a year) date of the discovery, namely 1669. There is no reliable information whether man had known phosphorus or its compounds before that or not. The unexpected discovery of phosphorus in the 17th Century profoundly impressed the academic world and was a real sensation owing to unusual property of the substance (it is too early to name it an "element"). It glowed in air at room temperature. Such compounds (for instance, Bologna stone - the product of Calcination of baryta with coal and oil, i.e. barium sulphide BaS), were called "phosphors" (from the Greek phos, light and phoro, to bear). Thus, the name appeared prior to the discovery of the element itself.

The history of its discovery was also unusual. There once lived in Hamburg a bankrupt merchant by the name of Hennig Brand. At that time, alchemy had already begun to lose ground but the belief in the "Philosophers' Stone" was still alive. H. Brand was one of those who believed in it. With a view to mending his business, he began to search for primary matter in various compounds. Human urine was one of the materials he analysed. H. Brand evaporated urine up to a syrupy liquid, distilled it and obtained a red liquid which he named urine oil. Having distilled this liquid once more, Brand saw a black precipitate at the bottom of his retort. After prolonged calcination the residue transformed into a white glowing substance precipitated on the walls of the vessel. Imagine the joy of the alchemist! He was sure that he had succeeded in isolating elementary fire. H. Brand tried to keep his discovery a secret and continued the work with phosphorus hoping to obtain gold from other metals. These efforts, as one might have expected, were in vain.

But H. Brand could not keep his secret for a long time and he finally revealed it himself. Having failed to obtain gold from other metals, Brand decided to put the new remarkable substance on sale keeping secret the method of its preparation. But in this attempt he also failed. As soon as phosphorus became known in Europe, it attracted attention of many scientists: the famous mathematician G. Leibniz, J. Kraft, J. Kunkel, R. Boyle, Ch. Huygens and many other chemists and physicists. J. Kunkel, who was at that time the alchemist at the court of the Prince of Saxony, sent J. Kraft, his assistant, to Hamburg to get the secret of phosphorus preparation from Brand. J. Kraft bought the secret for 200 thalers but it did not reach Kunkel. Kraft decided to keep the method of preparing the new substance to himself; he went on a trip of Europe to impress society with the marvellous substance's glow. J Kunkel tried to prepare phosphorus himself and after long work he succeeded in separating the new element.

The details of the method by which H. Brand prepared phosphorus did not reach us but the method of Kunkel (1676) is known rather well. Fresh urine was evaporated forming a black precipitate which was heated at first carefully and then intensively with sand and coal. After removal of volatile and oily compounds, phosphorus precipitated on cold walls of the retort as a white deposit. The following chemical reactions were involved in the process:

(a) $NaNH_4HPO_4 \xrightarrow{t} NaPO_3 + NH_3 \uparrow + H_2O$

b) $2NaPo_3 + Sio_2 \xrightarrow{t} Na_2Sio_3 + P_2O_5$

(c) $P_2O_5 + 5C \xrightarrow{t} P_2 + 5Co \uparrow$

However, Kunkel also decided not to make the method public. In 1680 R. Boyle became the third scientist to obtain phosphorus by approximately the same method; he reported it in a private letter to the London Royal Society. A. Hanckewitz, Boyle's assistant, organized production of phosphorus on a fairly large scale, deriving large profits since phosphorus was expensive.

It was believed for a long time that phosphorus existed only in one (white) allotropic modification but in 1847 A. Schroeter, heating white phosphorus up to 300°C without air, obtained red phosphorus, which, in contrast to the white phosphorus, was neither toxic nor combustible in air. In 1934 P. Bridgeman obtained the third modification, namely, black phosphorus, having subjected phosphorus to heating under high pressure.

ARSENIC

Arsenic - brittle, greyish-white, semi-metallic element (a metalloid), symbol As, atomic number 33, relative atomic mass 74.92. It occurs in many ores and occasionally in its elemental state and is widely distributed, being present in minute quantities in the soil, the sea and the human body. In larger quantities, it is poisonous. The chief source of arsenic compounds is as a by-product from metallurgical processes. It is used in making semiconductors, alloys and solders.

As it is a cumulative poison, its presence in food and drugs is very dangerous. The symptoms of arsenic poisoning are vomiting, diarrhoea, tingling and possibly numbness in the limbs and collapse. Its name derives from the Latin arsenicum.

Arsenic compounds, namely its sulphides AS_2S_3 (orpiment) and AS_4S_4 (realgar or sandarac), were well known to Greeks and Romans. Orpiment was also known under the name of "arsenic". Pliny the Elder and Dioscorides mentioned the toxicity of these compounds; Dioscorides noted calcination of "arsenic" to obtain white arsenic (oxide).

Arsenic is sometimes found in nature in native state and is fairly easily extracted from its compounds. It is not known who produced elemental arsenic for the first time. Usually, its discovery is ascribed to the alchemist Albert the Great. Paracelsus described the process of preparing metallic arsenic by the calcination of "arsenic" with egg-shells. According to some reports, metallic arsenic was known much earlier but it was considered to be a variety of native mercury. This is due to the fact that arsenic sulphide resembles one of mercury minerals and the extraction of arsenic from its ores is rather simple.

In the Middle Ages, arsenic was known not only in Europe but in Asia as well. Chinese alchemists could extract arsenic from its ores. Medieval Europeans had no way of knowing whether death of a person was caused by arsenic poisoning but

Chinese alchemists had a method of making sure. Unfortunately, their method of analysis is unknown. In Europe, the test for estimating arsenic content in human body and the food eaten before death was developed by D. Marsh. This test is very sensitive and is still used.

Since arsenic sometimes accompanies tin, there are reported cases (for instance, in Chinese literature) when people were poisoned by water or wine kept for some time in new tin vessels.

For a long time people confused white arsenic, or its oxide, with arsenic itself believing the two to be the same substance. The confusion was eliminated at first by H. Brand and then by A. Lavoisier who proved that arsenic is an independent chemical element.

Arsenic oxide has for a long time been used to kill rodents and insects. The symbol As originates from the Latin word arsenicum whose etymology is obscure.

ANTIMONY

Antimony - silver-white, brittle, semi-metallic element (a metalloid), symbol Sb (from Latin stibium), atomic number 51, relative atomic mass 121.75. It occurs chiefly as the ore stibnite and is used to make alloys harder; it is also used in photosensitive substances in colour photography, optical electronics, fireproofing, pigment and medicine. It was employed by the ancient Egyptians in a mixture to protect the eyes from flies.

Antimony and its compounds have been known from times immemorial. Some scholars say that metallic antimony was used in South Babylon for making vessels about 3400 years B.C. But in antiquity, antimony was mainly used for making cosmetics such as rouge and black paint for eye brows. In Egypt, however, antimony was apparently unknown or almost unknown. This is borne out by finds from Egyptian burial sites, particularly, by painted mummies.

In antiquity, antimony was confused with lead. It was only in alchemical literature of the Renaissance period that antimony was given a sufficiently accurate description. For example, G. Agricola clearly pointed out that antimony is a metal different from other metals. Basilius Valentinus devoted to antimony a whole treatise, Triumphal Carriage of Antimonium, in which he described the uses of antimony and its compounds.

There are several interpretations of the Latin name of antimony antimonium. Most likely it originates from the Greek word antimonos, which means "an enemy of solitude" and underlines simultaneous occurrence of antimony and other minerals.

BISMUTH

Bismuth - hard, brittle, pinkish-white, metallic element, symbol Bi, atomic number 83, relative atomic mass 208.98. It has the highest atomic number of all the stable elements (the elements from atomic number 84 up are radioactive). Bismuth occurs in ores and occasionally as a free metal (native metal). It is a poor conductor of heat and electricity and is used in alloys of low melting point and in medical compounds to soothe gastric ulcers. The name comes from the Latin besemutum, from the earlier German Wismut (of unknown origin).

Bismuth has been known to mankind for centuries but for a long time it was confused with antimony, lead and tin. Paracelsus, for instance, said that there were two varieties of antimony - a black one used for the purification of gold and very similar to lead and a white one named bismuth and resembling tin; a mixture of these two varieties resembles silver. From the chemical standpoint, this confusion can easily be explained. Antimony and bismuth are analogues of each other and have common features with lead and tin, the elements of the previous group.

Agricola, unlike Paracelsus, gave a rather detailed description of bismuth and of the process of its extraction from ores mined in Saxony. Miners thought that bismuth, as well as tin, were a variety of lead and that bismuth could be transformed into silver.

In Central Russia, bismuth has been known since the 15th century. With the development of book printing bismuth, along with antimony, began to be used for casting typographical types. In literature, few elements have such a great number of names as bismuth. E. Von Lippmann in his book History of Bismuth from 1480 to 1800 gives twenty one names of this metal used in Europe. A sufficiently clear idea of bismuth as an independent metal was formed only in the 18th century.

ZINC

Zinc (Germanic Zint 'point') - hard, brittle, bluish-white, metallic element, symbol Zn, atomic number 30, relative atomic mass 65.37. The principal ore is sphalerite or zinc blende (zinc sulphide, ZnS). Zinc is little affected by air or moisture at ordinary temperatures; its chief uses are in alloys such as brass and in coating metals (for example, galvanized iron). Its compounds include zinc oxide, used in ointments (as an astringent) and cosmetics, paints, glass and printing ink.

Zinc has been used as a component of brass since the Bronze Age, but it was not recognized as a separate metal until 1746, when it was described by German chemist Andreas Sigismund Marggraf (1709-1782). The name derives from the shape of the crystals on smelting.

The zinc industry in Europe generates about 80000 tons of zinc waste each year.

Zinc is also one of the elements whose compounds have been known to mankind from time immemorial. Its best known mineral was calamine (zinc carbonate). Upon calcination it yielded zinc oxide, which was widely used, for instance, for treating eye diseases.

Although zinc oxide is comparatively easily reduced to free metal, it was obtained in a metal state much later than copper, iron, tin, and lead. The explanation is that reduction of zinc oxide with coal requires high temperature (about 1100°C). The boiling point of the metal is 906°C; therefore, highly volatile zinc vapour escapes from the reaction zone.

Before metallic zinc was isolated, its ores were used for making brass, an alloy of zinc and copper. Brass was known in Greece, Rome, India, and China. It is an established fact that Romans produced brass for the first time during the reign of Augustus (B.C. 20-A.D. 14). Interestingly, the Roman method of preparing brass was still used up to the 19th century.

It is impossible to establish when metallic zinc was obtained. In ancient Dacian ruins an idol was found containing 27.5 percent of zinc. Zinc was possibly obtained during brass production as a side product.

In the 10-11th centuries, the secret of zinc production was lost in Europe and zinc had to be imported from India and China. It is believed that China was the first country to produce zinc on a large scale. The production process was extremely simple. Earthenware filled with calamine were tightly closed and piled into a pyramid. The gaps between the pots were filled with coal and the pots were heated to red heat. After cooling the pots, where zinc vapours condensed, were broken and metal ingots were extracted.

Europeans rediscovered the secret of zinc production in the 16th century when zinc had already been recognized as an independent metal. During the next two centuries, many chemists and metallurgists worked on with methods of zinc extraction. A great deal of credit should go to A. Marggraf who published in 1746 a large treatise, Methods of Extraction of Zinc from its Native Mineral Calamine. He also found that lead ores from Rammelsberg (Germany) contained zinc and that zinc could be obtained from sphalerite, natural zinc sulphide.

The name "Zinc" originates from the Latin word denoting leucoma or white deposit. Some scholars relate "Zinc" to the German word zink, which means lead.

3

Elements of air and water

This chapter is devoted to three elemental gases–hydrogen, nitrogen and oxygen – whose discovery was the most important event in chemistry of the second half of the 18th century. Nitrogen and oxygen constitute almost the whole of the earth's atmosphere, other gases being only present in low concentrations. Hydrogen and oxygen form water – one of the most amazing compounds. All three elements together with carbon comprise organic compounds and are found in all animals and plants without exception.

Discovery of hydrogen, nitrogen and oxygen and their proper understanding played an extremely important role in the development of chemistry since it contributed to the emergence of many modern concepts. Here is a short list of achievements directly related to the discovery of these gases: the oxygen theory of combustion (A. Lavoisier); the atomistic theory (J. Dalton); the theory of acids and bases; the use of oxygen and hydrogen scales of atomic weights (masses); conception of hydrogen as primary matter which gave rise to all other elements (V. Prout).

The discoveries of hydrogen, nitrogen and oxygen occupy a special place in the history of elements. The understanding of the real nature of these elements was a complex, contradictory and prolonged process. Discovering new gaseous products in the course of chemical reactions (hydrogen, nitrogen and oxygen), scientists did not know yet that they were dealing with new chemical elements.

From time immemorial only one type of gas, namely air, was known; it was studied by physics and was not in chemistry's sphere of interests. The gaseous products that were formed during various processes (for instance, fermentation or putrefaction) were considered by scientists to be varieties of air. The concept of "gas" appeared only at the beginning of the 17th century. It was introduced by J. Van Helmont, a famous natural scientist. He derived it from the Greek word chaos. Once J. Van Helmont burnt 62 pounds of wood and obtained only one pound of ash. What was the rest of the wood transformed into? Into "wood spirit" (spiritus silvester), the scientist believed. He wrote that he called this previously unknown "Spirit" by a new name "gas". Now we know that the scientist obtained carbon dioxide which was produced again by the English physicist J. Black only over 100 years later. But J. Van Helmont did not understand his discovery: he saw in the "wood spirit" only a variety of air.

Therefore, we have no right to apply the term "the discovery of a new element" in its latter-day sense to the constituents of air and water. On the other hand, the discoveries of hydrogen, nitrogen and oxygen differ considerably from those chance discoveries that had taken place in the pre-scientific period. Firstly, in the 18th century there was a well-developed theory named "the theory of phlogiston" (the phlogistic theory). Secondly, the gaseous state of matter became at last, owing to J. Van Helmont, the subject of chemical study and to a new branch of chemistry – pneumatic chemistry was born with its own research methods and laboratory equipment. In other words, the discovery of elemental gases became possible due to purposeful experimental work based on theoretical conceptions. And before we begin the story of these elements, we have to consider the phlogistic theory and pneumatic chemistry.

In essence, the phlogistic theory was very simple and therefore, seemed to be very convincing. Its name originates from the Greek word phlogistos, which means "combustible". The theory provided an explanation of processes taking place during combustion, calcination of metals and respiration, the essence of which was unclear. So the idea of a substance which is the main participant in all the above processes – phlogiston – was put forward.

Although ideas about materia ignea were expressed in one form or another by several scientists, the German chemist and physician G. Stahl is regarded as the true founder of the phlogistic theory. He reasoned in the following way. All bodies can burn only owing to the presence of phlogiston in them. The more phlogiston a body contains, the more actively it burns. Coal is an example of a substance which is almost pure phlogiston. Upon calcination, metals lose phlogiston and transform into "calx" (earths). The addition of phlogiston to calcinated metal produces pure metal again. Calcination of metal scale with coal is a good illustration. This process was well known even to primitive metallurgists.

From the standpoint of modern chemistry, all this means that in the course of an oxidation reaction (for instance, the formation of an oxide during calcination of a metal) phlogiston is lost; on the contrary, in a reduction reaction (calcination of metal oxide with coal) phlogiston is acquired. Everything is so simple and clear. But even a beginner in chemistry will understand that the phlogistic theory is erroneous. It follows from this theory that the weight of the substance upon combustion must decrease rather than increase; a metal oxide must be lighter than the metal itself. According to the phlogistic theory, metals should be considered as complex compounds (metal plus phlogiston) and their oxides (earths) as simple substances (metal minus phlogiston).

And, nevertheless, the phlogistic theory was recognized for about a century and was earnestly advocated by famous chemists of that time including G. Canvendish, J. Priestley and C. Scheele whose names are associated with the discovery of the elements of air and water. At the initial stages of their discoveries, the concepts of the phlogistic theory played an important role.

New interest in the study of gases contributed to the development of pneumatic chemistry and it was the second inevitable step towards the discovery of hydrogen, nitrogen and oxygen. The study of gases had for a long time been made difficult by the absence of adequate methods for their preparation and collection and analysis of their properties. Bladders of animals were almost the only experimental vessels for collecting and weighing the liberated gases. It proved much more difficult to study gases than solids or liquids.

At the beginning of the 18th century S. Hales, an English scientist, invented a pneumatic bath. In this apparatus, the vessel where a gas was formed (a retort with a reaction mixture) was separated from the collector for the liberated gas. The collector was a flask which was turned upside down and filled with water. Penetrating into the flask, the gas bubbles displaced water and the flask became filled with the gas under study.

J. Black (an English scientist, one of the founders of pneumatic chemistry) also made use of the pneumatic bath to study compounds known for a very long time - lime and magnesia alba (calcium and magnesium carbonates). Their calcination or reaction with acids produces a gas. Now we can easily guess that this gas was the same "wood spirit" J. Van Helmont had obtained by burning charcoal. Helmont, however, did not go beyond establishing the fact and offering some vague speculations. Black advanced much further. He noticed that compounds formed upon calcination or in the reactions with acids can be transformed into the initial state.

Now a chemist would comment upon this achievement in the following way: the scientist carried out a forward reaction (decomposition of carbonates into oxides and carbon dioxide) and a reverse reaction (addition of carbon dioxide to the oxides yielding the initial product). The mass of the initial products was completely restored and thus, J. Black succeeded in what others had failed.

He weighed some gas in a bound state, referring to it as "bound" or "fixed" air. The gas was liberated during fermentation processes or combustion of charcoal but it did not sustain respiration or combustion. Black believed this gas to be an independent constituent of atmospheric air.

Thus, in 1754 carbon dioxide was discovered under the name of "fixed" air. This event was of an extreme importance for the subsequent discovery of other gases mainly because of the fact that after inevitable arguments and discussions the scientists began to consider carbon dioxide to be not a variety of air but an independent substance different from air and contained in many solids. And since on addition of carbon dioxide to oxides the mass of the product formed exceeded that of the initial product, the main principle of the phlogistic theory was undermined. It was, however, a long time before the significance of this fact was recognized and the phlogistic theory ceased to be the only basis for the explanation of many observations of pneumatic chemistry.

HYDROGEN

Hydrogen (Greek hydro+gen "water generator") - colourless, odourless, gaseous, non-metallic element, symbol H, atomic number 1, relative atomic mass 1.00797. It is the lightest of all the elements and occurs on Earth chiefly in combination with oxygen as water. Hydrogen is the most abundant element in the universe, where it accounts for 93% of the total number of atoms and 76% of the total mass. It is a component of most stars, including the sun, whose heat and light are produced through the nuclear-fusion process that converts hydrogen into helium. When subjected to a pressure 500000 times greater than that of the Earth's atmosphere, hydrogen becomes a solid with metallic properties, as in one of the inner zones of Jupiter. Hydrogen's common and industrial uses include the hardening of oils and fats by hydrogenation, the creation of high-temperature flames for welding and as rocket fuel. It has been proposed as a fuel for road vehicles.

Its isotopes deuterium and tritium (half-life 12.5 years) are used in nuclear weapons and deuterons (deuterium nuclei) are used in synthesizing elements. The element's name refers to the generation of water by the combustion of hydrogen and was coined in 1787 by French chemist Louis Guyton de Morveau (1737-1816).

Hydrogen is one of the most striking elements of the periodic system, it is number one and the lightest of all the existing gases. It is the element whose discovery was indispensable for the solution of many problems of chemical theory. It is an element whose atom, losing its only valence electron, becomes a "bare" proton. And, therefore, chemistry of hydrogen is, in a way, unique; it is the chemistry of an elementary particle.

Once D.I. Mendeleev called hydrogen the most typical of typical elements (meaning the elements of the short periods in the system), because it begins the natural series of chemical elements.

And such a fascinating element is readily available: it can be obtained without difficulty in any school laboratory, for instance, by pouring hydrochloric acid on zinc shavings.

Even in those bygone times, when chemistry was not a science yet and when alchemists were still searching for the "philosophers' stone", hydrochloric, sulphuric and nitric acids as well as iron and zinc were already known. In other words, man had in his possession all components whose reaction could give rise to hydrogen. Only a chance was needed and chemical literature of the 16-18th centuries reported that many times chemists observed how the pouring of, for instance, sulphuric acid on iron shavings produced bubbles of a gas which they believed to be an inflammable variety of air.

One of those who observed this mysterious variety of air was the famous Russian scientist M.V. Lomonosov. In 1745 he wrote a thesis, On Metallic Lustre, which said, among other things: "On dissolution of some base metal, especially iron, in acidic alcohols, inflammable vapour shots out from the opening of the flask." (According to the terminology of those times, acidic alcohols meant acids.) Thus, M.V. Lomonosov observed none other than hydrogen. But the sentence went on to read: "which is phlogiston." Since metal dissolved in the acid liberating materia ignea or "inflammable vapour", it was very convenient to assume that dissolving metal releases phlogiston: everything fits nicely into the theory of phlogiston.

And now is the time to meet the outstanding English Scientist H. Cavendish, a man fanatically devoted to Science and an excellent experimenter. He never hurried with making public his experimental results and sometimes several years had to pass before his articles appeared. Therefore, it is difficult to pinpoint the date when the scientist observed and described the liberation of "inflammable air". What is known is that this work published in 1766 and titled "Experiments with Artificial Air" was done as a part of pneumatic chemistry research. It is also likely that the work was performed under the influence of J. Black. H. Cavendish had become interested in fixed air and decided to see whether there existed other types of artificial air.

In this manner, the scientist referred to the variety of air which is contained in compounds in a bound state and which can be separated from them artificially. H. Cavendish knew that inflammable air had been observed many times. He himself obtained it by the same technique: the action of sulphuric and hydrochloric acids on iron, zinc and tin, but he was the first to obtain definite proof that the same type of air was formed in all cases – inflammable air. And he was the first to notice the unusual properties of inflammable air. As a follower of the phlogistic theory, H. Cavendish could give only one interpretation of the substance's nature. Like M.V. Lomonosov, he identified it as phlogiston. Studying the properties of inflammable air, he was sure that he was studying the properties of phlogiston. H. Cavendish believed that different metals contain different proportions of inflammable air. Thus, to the fixed air of J. Black, the inflammable air of H. Cavendish was added. Strictly speaking, the two scientists discovered nothing new: each of them only summarized the data of previous observations. But this summing up represented considerable progress in the history of human knowledge.

Fixed air and inflammable air differed both from ordinary air and from each other. Inflammable air was surprisingly light. H. Cavendish found that phlogiston, which he had separated, had a mass. He was the first to introduce a quantity to characterize gases, that of density. Having assumed the density of air to be unity, Cavendish obtained the density of 0.09 for inflammable air and 1.57 for fixed air. But here a contradiction arose between Cavendish the experimenter and Cavendish the adherent of the phlogistic theory. Since inflammable air had a positive mass, it could by no means be considered to be pure phlogiston. Otherwise, metals losing inflammable air would have to lose mass as well. To avoid the contradiction, Cavendish proposed an original hypothesis: inflammable air is a combination of phlogiston and water. The essence of the hypothesis was that at last hydrogen appeared in the composition of inflammable air.

The evident conclusion is that Cavendish, like his predecessors, did not understand the nature of inflammable air, although he had weighed it, described its properties, and considered it to be an independent kind of artificial air. In a word, Cavendish, unaware of the fact, studied "phlogiston" obtained by him as he would have studied a new chemical element. But Cavendish could not perceive that inflammable air was a gaseous chemical element – so strong were the chains of the phlogistic theory. And having found that the real properties of inflammable air contradicted this theory, he came up with a new hypothesis, as erroneous as the theory itself.

Therefore, strictly speaking, the phrase "hydrogen was discovered in 1766 by the English scientist H. Cavendish" is meaningless. Cavendish described the processes of preparation and the properties of inflammable air in greater detail than his predecessors. However, he "knew not what he was doing". The elementary nature of inflammable air remained beyond his grasp. It was not the scientist's fault, however; chemistry had not yet matured enough for such an insight. Many years passed before hydrogen became, at last, hydrogen and occupied its proper place in chemistry.

Its Latin name hydrogenium stems from the Greek words hydr and gennao which mean "producing water". The name was proposed in 1779 by A. Lavoisier after the composition of water had been established. The symbol H was proposed by J. Berzelius.

Hydrogen is a unique element in the sense that its isotopes differ in their physical and chemical properties. At one time, this difference prompted some scientists to consider hydrogen isotopes as independent elements and to find for them special boxes in the periodic table. Therefore, the history of the discovery of hydrogen isotopes is of special interest, as a continuation of the history of hydrogen itself.

The search for hydrogen isotopes began in the twenties of this century but all attempts were unsuccessful, resulting in the belief that hydrogen had no isotopes. In 1931, it was suggested that hydrogen, nevertheless, contains a heavy isotope with a mass number of 2. Since this isotope had to be twice as heavy as hydrogen, the scientists tried to isolate heavy hydrogen by physical methods. In 1932 the American scientists Urey, Brickwedde and Murphy evaporated liquid hydrogen and studying the residue by spectroscopy, found a heavy isotope in it. In the atmosphere it was discovered only in 1941. The name "deuterium" originates from the Greek word deuteros which means "second, another one". The next isotope with a mass number of 3, tritium (from the Greek tritos – the third), is radioactive and was discovered in 1934 by English scientists M. Oliphant, P. Hartec and E. Rutherford. The name "protium" was assigned to the main hydrogen isotope. This is the only case when isotopes of the same element have different names and symbols (H, D, and T). 99.99 percent of all hydrogen is protium; the rest is deuterium with only traces of tritium.

NITROGEN

Nitrogen (Greek nitron "native soda", Sodium or potassium nitrate) - colourless, odourless, tasteless, gaseous, non-metallic element, symbol N, atomic number 7, relative atomic mass 14.0067. It forms almost 80% of the Earth's atmosphere by volume and is a constituent of all plant and animal tissues (in proteins and nucleic acids). Nitrogen is obtained for industrial use by the liquefaction and fractional distillation of air. Its compounds are used in the manufacture of foods, drugs, fertilizers, dyes and explosives.

Nitrogen has been recognized as a plant nutrient, found in manures and other organic matter, from early times, long before the complex cycle of nitrogen fixation was understood. It was isolated in 1772 by English chemist Daniel Rutherford (1749-1819) and named in 1790 by French chemist Jean Chaptal (1756-1832).

Nitrogen is used in the Haber process to make ammonia, NH_3 and to provide an inert atmosphere for certain chemical reactions.

Although fixed air (carbon dioxide) and inflammable air (hydrogen) were later found in the earth's atmosphere, their discoveries did not result from the study of atmospheric air. The latter was still regarded as "classical" air and nobody had any idea that it was a mixture of gases. It was the study of atmospheric air, however, that made it possible for pneumatic chemistry to obtain the most valuable results.

The study of the atmosphere led to the discovery of nitrogen. Although it is associated with the name of a certain scientist and a certain date, this certainty is misleading. It is rather difficult to separate the history of nitrogen discovery from the mainstream of pneumatic chemistry; one can only think of a more or less logical sequence of events.

Very early in history man came across nitrogen compounds, for instance, saltpetre and nitric acid, frequently observing liberation of brown vapours of nitrogen dioxide. Obviously, it would be impossible to discover nitrogen by decomposing its inorganic compounds. Tasteless, colourless, odourless and chemically rather inactive, nitrogen would have remained unnoticed.

Therefore, it is not easy to decide where to begin the story of the discovery of nitrogen. Although our choice may seem subjective, we start with 1767 when H. Cavendish and J. Priestley, another outstanding English physicist, chemist and philosopher, set out to study the action of electric discharges on various gases. There were only few such gases at that time: ordinary air, fixed air and inflammable air. Although the experiments did not produce definite results, it was shown later that electric discharge in humid air yields nitric acid. Later this fact proved to be useful for the analysis of the earth's atmosphere.

In 1777, H. Cavendish reported in a private letter to J. Priestley that he had succeeded in preparing a new variety of air named by him asphyxiating or mephitic air. Cavendish repeatedly passed atmospheric air over red-hot coal. The resulting fixed air was absorbed with alkali. The residue was mephitic gas. Cavendish did not study it thoroughly and only reported the fact to Priestley. Cavendish returned to the study of mephitic air much later, did a large work but the credit for the discovery had already gone to another scientist.

When Priestley received the letter from Cavendish he was busy with important experiments and read it without due attention. Priestley burned various inflammable compounds in a given volume of air and calcinated metals; the fixed air formed during these processes was removed with the aid of limewater. The main thing which he noticed was that the volume of air decreased considerably. A reader will prompt that as a result of metal calcination or combustion of compounds, the oxygen present in the apparatus was bonded and nitrogen remained. Priestley, however, had no idea about the existence of such a gas as oxygen (two years later, however, he became one of its discoverers) and to explain the observed phenomenon, he turned to phlogiston. Priestley believed that the result of metal calcination was due exclusively to the action of phlogiston. The remaining air is saturated with phlogiston and consequently, it can be named "phlogisticated" air; it sustains neither respiration nor combustion.

Thus, Priestley was in possession of a gas which subsequently became known as nitrogen. But this extremely important result was treated by him as something of secondary importance. The existence of "phlogisticated" air was for Priestley evidence of the fact that phlogiston did play a role in natural processes. This story shows once more how the erroneous phlogistic theory hampered the discovery of elemental gases.

So, neither Cavendish nor Priestley could understand the real nature of the new gas. The understanding came later when oxygen appeared on the scene of chemistry. English physician D. Rutherford, the pupil of J. Black, who is considered to be the discoverer of nitrogen, did, in principle, nothing new compared with his famous colleagues. In September 1772, Rutherford published a magisterial thesis on the so-called Fixed and Mephitic Air in which he described the properties of nitrogen. This gas, according to Rutherford, was absorbed neither by limewater nor by alkali and was unsuitable for respiration; he named it "corrupted" air.

Not properly discovered or understood as a gaseous chemical element, nitrogen in the seventies of the 18th century had three names which confused still more the fuzzy chemical concept muddled by the persisting influence of the phlogistic theory. Phlogisticated, mephitic, or corrupted air was yet to receive its final name.

This name was proposed in 1787 by A. Lavoisier and other French Scientists who developed the principles of a new chemical nomenclature. They derived the word "azote" from the Greek negative prefix "a" and the word "zoe" meaning "life". Lifeless, not supporting respiration and combustion, that was how the chemists saw the main property of nitrogen. Later this view turned out to be erroneous: nitrogen is vitally important for plants. The name "azote", however, remained. The symbol of the element, N, originates from the Latin nitrogenium which means "saltpeter-forming".

H. Cavendish studied the properties of nitrogen in detail. He was one of the first scientists to believe that phlogisticated air is a component of ordinary air. One day, in the course of his experiments Cavendish questioned the homogeneity of phlogisticated air. He passed an electric spark through its mixture with oxygen transforming the whole into nitrogen oxides which were removed from the reaction zone. But every time a small fraction of the phlogisticated air (nitrogen) remained unchanged and did not react with oxygen. It was a very small fraction, a slightly noticeable gas bubble - only 1/125 of all nitrogen taken for the experiment. Cavendish could not explain the reason for this phenomenon observed in 1785. The answer was found only over one hundred years later. You will read about it in chapter 9 devoted to inert gases.

OXYGEN

Oxygen (Greek oxys "acid" genes "forming") - colourless, odourless, tasteless, non-metallic, gaseous element, symbol O, atomic number 8, relative atomic mass 15.9994. It is the most abundant element in the Earth's crust (almost 50% by mass), forms about 21% by volume of the atmosphere and is present in combined form in water and many other substances. Life on Earth evolved using oxygen, which is a by-product of photosynthesis and the basis for respiration in plants and animals.

Oxygen is very reactive and combines with all other elements except the inert gases and fluorine. It is present in carbon dioxide, silicon dioxide (quartz), iron ore, calcium carbonate (limestone). In nature it exists as a molecule composed of two atoms (O_2); single atoms of oxygen are very short-lived owing to their reactivity. They can be produced in electric sparks and by the sun's ultraviolet radiation in space, where they rapidly combine with molecular oxygen to form ozone (an allotrope of oxygen).

Oxygen is obtained for industrial use by the fractional distillation of liquid air, by the electrolysis of water, or by heating manganese oxide with potassium chlorate. It is essential for combustion and is used with ethyne (acetylene) in high-temperature oxyacetylene welding and cutting torches.

The element was first identified by English chemist Joseph Priestley in 1774 and independently in the same year by Swedish chemist Karl Scheele. It was named by French chemist Antoine Lavoisier in 1777.

One can safely say that none of the chemical elements played such an important role in the development of chemistry as oxygen. This life-giving gas enabled chemistry to make such great progress at the end of the 18th century which had never been possible before. First of all, oxygen overturned the phlogistic theory which had seemed immovable. Erroneous as it was, this theory was undoubtedly of some historical usefulness. For the time being, the theory of phlogiston made it possible somehow to systematize the existing chemical conceptions and to consider various processes in nature and laboratory from a common (though erroneous) standpoint. This gave a certain purposefulness to the research. Hydrogen and nitrogen were found from the phlogistic conceptions but the study of these "varieties of air" made it possible to accumulate new facts whose explanation demanded a different approach. Figuratively speaking, chemistry needed a new look at itself and oxygen made it possible.

In defiance of the theory of phlogiston, vague conjectures were repeatedly made that combustion of inflammable compounds and calcination of metals drew a "substance" from the air. In 1673 R. Boyle concluded that when lead and antimony are calcinated, a very fine "material ignea" passes into the metals and combines with them, increasing their weight. "...the good Robert Boyle's opinion is false," Lomonosov wrote 80 years later. The famous Russian scientist said that air participates in the processes of combustion - particles from the air mix with the compound being calcinated and increase its mass.

In the period when pneumatic chemistry was gaining ground, the French chemist P. Bayen wrote a paper (1774) in which he discussed the causes for an increase in the mass of metals during calcination. He believed that a peculiar variety of air - an expansible fluid, heavier than ordinary air - was added to a metal in the process of calcination. Bayen obtained this fluid by thermal decomposition of mercury compounds. And, conversely, acting on metallic mercury, the fluid transformed it into a red compound.

Bayen, unfortunately, only established the facts and did not pursue the subject further. However, you will see later that the scientist was actually dealing with oxygen. Remember two things: the date 1774 and the compound observed by Bayen - red mercury oxide. In the same year J. Priestley experimented with the same compound. Shortly before, he had discovered that in the presence of green plants fixed air, not supporting respiration, turned into ordinary air suitable for respiration by living organisms. This fact was extremely important not only for chemistry but for biology as well. Priestley proved for the first time that plants release oxygen.

In the early 1770's so-called Saltpetre gas was already known. It was liberated in the reaction of diluted nitric acid with iron shavings (it is nitrogen oxide in modern terminology). It turned out that saltpetre gas can be transformed (by its reaction with iron dust) into another variety of air supporting combustion but not supporting respiration. Thus, J. Priestley discovered another nitrogen oxide, N_2O and named it, according to the logic of the phlogistic theory, dephlogisticated saltpetre gas.

August 1, 1774, which was to become a milestone in the history of chemistry, was a usual day of hard work for J. Priestley. He placed red mercury oxide into a sealed vessel and directed on it sunbeams, focused with a big lens. The compound began to decompose yielding bright metallic mercury and a gas (several years later this gas would be named "Oxygen" and become the third elemental gas). Unlike nitrogen, oxygen was not initially isolated from the atmosphere. The new air variety was extracted from a solid. The gas discovered by Priestley proved to be suitable for respiration. A candle burnt in the atmosphere of this gas much brighter than in ordinary air. Nothing was observed when the new gas was mixed with air but, being mixed with saltpetre, the gas yielded brownish vapours (known at present to be NO_2 formed from NO). A similar, although not so pronounced, picture was observed when saltpetre was reacted with ordinary air. Priestley had only to say: "The new gas is a component of air." But he was not ready yet to do so and named the new variety of air "dephlogisticated" air - something quite natural for a follower of the phlogistic theory.

After the discovery - and this is an important detail in the history of oxygen - Priestley left for Paris where he told Lavoisier and some other French scientists about his experiments. Lavoisier appreciated at once the importance of the experiment of his English colleague - he had a much clearer idea about it than Priestley. But Priestley kept talking about dephlogisticated air, still in the grip of his delusion (which is another proof of the vitality of the phlogistic theory). Unable to see the greatness of his own discovery, Priestley considered dephlogisticated air to be a complex substance. Only in 1786, influenced by the ideas of Lavoisier, did he begin to view it as an elemental gas.

Thus, we owe the discovery of oxygen to P. Bayen and J. Priestley. However, a third name should he added that of the famous Swedish chemist C. Scheele. It became widely known to the scientific community when Scheele published the book chemical treatise about Air and Fire. Written in 1775, it appeared only two years later for which the publisher was to blame. This disappointing fact deprived Scheele of the right to be named the discoverer of oxygen although he had described it as early as 1772 and his description was much more detailed and accurate than that of Bayen and Priestley. Scheele obtained oxygen ("fiery air") in various ways, by decomposing inorganic compounds, distillation of saltpetre with sulphuric acid yielded brown vapours which became colourless at high temperatures. Scheele collected these vapours and named the new gas "fiery air". In this gas, like in Priestley's, a candle burned much brighter than in ordinary air. Scheele believed that fiery air was a component of ordinary air. Mixing it with mephitic or corrupted air of Rutherford, Scheele prepared a mixture which did not differ at all from ordinary air. In fact, the scientist realized that atmospheric air is a mixture of gases which later were to be known as nitrogen and oxygen. However, this seems to be easy only with our superior knowledge. Scheele was deluded by his devotion to the phlogistic theory. Burning inflammable air (hydrogen) in a vessel with air, the scientist did not detect any products of the reaction of inflammable air with fiery one. His conclusion was that this reaction produced heat. Scheele reasoned that fiery air, combining with phlogiston, produces heat (which had, according to Scheele, a material nature) whose "decomposition" yields fiery air.

Scheele discovered fiery air knowing nothing about Priestley's experiments and informed Lavoisier about it on September 30, 1774. Regretfully, Scheele's results were published too late. Had they appeared earlier, the difficult and contradictory process of elucidating the nature of elemental gases would have been accelerated.

Their real understanding was made possible by Lavoisier, one of the most outstanding chemists of all times. During the period dominated by the phlogistic theory, vast experimental material was accumulated which led to revolutionary changes in chemistry. The main credit for this goes to A. Lavoisier owing to whom oxygen was properly understood. F. Engels wrote: "Lavoisier was able to discover in the oxygen obtained by Priestley the real antipode to the fantastic phlogiston and thus could throw overboard the entire phlogistic theory. But this did not in the least do away with the experimental results of phlogistics. On the contrary, they persisted, only their formulation was inverted, was translated from the phlogistic into the now valid chemical language and thus they retained their validity."

Lavoisier's road to the discovery of oxygen was much straighter than that of his contemporaries. At first the French scientist also appealed to the phlogistic theory, but the more experimental facts he obtained, the more inclined he became to discard it. By November 1, 1772, he had finished the description of his experiments on the combustion of various compounds in air. He concluded that the mass of all substances, including metals, increases upon combustion and calcination.

Since these processes require a great amount of air, A. Lavoisier made another conclusion: Air is a mixture of gases with different properties. A certain part of it supports combustion and becomes bonded to the burning substance. At first A. Lavoisier assumed that this type of air is similar to the fixed air of J. Black but soon he saw his error. In February 1774, the French scientist discovered that air which interacts with a substance during combustion is most suitable for respiration. Thus, A. Lavoisier met face to face with oxygen but refrained from announcing the discovery of a new gas since he was going to perform some additional experiments.

In October 1774, Priestley reported to Lavoisier about his discovery revealing to the French chemist the real significance of his own findings. He immediately began to experiment with red mercury oxide which was the most suitable "generator" of oxygen. In April 1775, Lavoisier made a report to the Academy of Sciences: "Memoir on the Nature of the substance which combines with Metals upon calcination and Increases their weight" - the announcement of the discovery of oxygen.

Lavoisier wrote that this type of air had been discovered almost simultaneously be Priestley, Scheele and by him. At first he said that it was "very easily inhaled air" but then changed the name to "life-giving or invigorating" air.

This fact alone shows how far behind Lavoisier left his contemporaries in understanding the nature of oxygen. Invigorating air became the subject of comprehensive studies. At a later stage the scientist came to the conclusion that "the most easily inhaled air" is an acid-forming principle, the most important part of all acids. Later, it was shown that this belief was erroneous (when oxygen-free acids were described with hydrohalic acids as an example). But in 1779, Lavoisier thought it possible to reflect this property of the new gas in its name of "oxygen" derived from the Greek for "producing acid".

Determination of the water composition became a major advance of the oxygen theory. In 1781 H. Cavendish observed that inflammable air upon combustion is transformed almost completely (together with dephlogisticated air) into pure water. But he published his results only in 1784. Lavoisier knew about these experiments and after repeating them, he concluded that water is not a simple substance but a mixture of inflammable and invigorating air. Since the conclusion was made in 1783, Lavoisier is held by many to be the first one to have established the composition of water. In reality, however, H. Cavendish was the first. Determination of the composition of water made it possible to get an insight into the nature of hydrogen.

What makes the history of the discovery of oxygen interesting is that the process was not a single event. Several stages were passed: from an empirical observation of oxygen to a proper understanding of its nature as a gaseous chemical element. It should also be mentioned that the discovery of oxygen (as well as of other elemental gases) was not the doing of one man. Engels wrote: "Priestley and Scheele had produced

oxygen without knowing what they had laid their hands on...
And although Lavoisier did not produce oxygen simultaneously
and independently of the other two, as he claimed later on, he
nevertheless is the real discoverer of oxygen vis-a-vis the others
who had only produced it without knowing what they had
produced."

4

Elements discovered by chemical analysis

This chapter is devoted to the description of the elements that were discovered owing to chemical analysis of natural substances, mainly, various minerals. With progress of chemistry, its role in the study of inorganic nature was becoming more and more important. The chapter begins with the discovery of cobalt and ends with the discovery of vanadium. It covers the time period of about 100 years (from 1735 to 1830). During this period, more than 30 chemical elements were discovered due to the development of the chemical analysis. Of course, analysis played an important role in the discovery of some other elements as well, for instance, of rare earth elements, but because of the specific histories of these elements they will be discussed in a separate chapter.

COBALT

Cobalt (German kobalt "goblin") - hard, lustrous, grey, metallic element, symbol Co, atomic number 27, relative atomic mass

58.933. It is found in various ores and occasionally as a free metal, sometimes in metallic meteorite fragments. It is used in the preparation of magnetic, wear-resistant and high-strength alloys; its compounds are used in inks, paints and varnishes.

The isotope Co-60 is radioactive (half-life 5.3 years) and is produced in large amounts for use as a source of gamma rays in industrial radiography, research and cancer therapy. Cobalt was named in 1730 by Swedish chemist Georg Brandt (1694-1768); the name derives from the fact that miners considered its ore worthless because of its arsenic content.

The history of the discovery of cobalt can conveniently be started with the history of its name. Cobalt owes its name to the mineral which medieval Saxony miners named "cobold" after the evil spirit who was assumed to inhabit the mineral. This mineral closely resembled silver ore but all attempts to produce silver from it were unsuccessful.

Blue cobalt glasses, enamels and pigments were known as early as 5000 years ago in ancient Egypt. In Pharaoh Tutankhamen's tomb, archaeologists found fragments of blue glass. It is not known whether the preparation of blue glasses and paints on the basis of cobalt compounds was due to chance or whether it was a conscious effort. At any rate the method of their preparation remained unknown for a long time. Its first mention dates back to 1679.

Cobalt was discovered by the Swedish chemist G. Brandt in 1735. In his "Dissertation on semi-metals" Brandt wrote about a new semi-metal, cobold, discovered by him. By semi-metals the scientist meant compounds whose properties resemble those of known metals but which are not malleable. He described six semi-metals: mercury, bismuth, zinc, antimony, cobalt and arsenic. Since the majority of bismuth ores contain cobalt, G. Brandt proposed several methods to distinguish between cobalt and bismuth.

In 1744 G. Brandt found a new mineral which contained cobalt, iron and sulphur. It proved to be cobalt sulphide Co_3S_4.

Later, G. Brandt studied cobalt in detail. At the turn of the 18th century, cobalt and its compounds were studied by T. Bergman, L. Thenard, L. Proust and J. Berzelius, which made cobalt a well-investigated element. It must be added that for a long time many chemists did not believe in the discovery of cobalt. In 1776 the Hungarian chemist P. Padaxe said that cobalt was a compound of iron and arsenic; but he considered nickel, which had already been discovered by that time, to be a chemical element. Only by the end of the 18th century, the efforts of many scientists confirmed the discovery of G. Brandt.

Cobalt, as well as nickel, is often present (and sometimes in great amounts) in meteorites. In 1819 the German chemist F. Stromeyer reported the discovery of cobalt in a meteorite and somewhat later S. Tennant found nickel in the same meteorite.

NICKEL

Nickel - hard, malleable and ductile, silver-white metallic element, symbol Ni, atomic number 28, relative atomic mass 58.71. It occurs in igneous rocks and as a free metal (native metal), occasionally occurring in fragments of iron-nickel meteorites. It is a component of the Earth's core, which is held to consist principally of iron with some nickel. It has a high melting point, low electrical and thermal conductivity and can be magnetized. It does not tarnish and therefore is much used for alloys, electroplating and for coinage.

It was discovered in 1751 by Swedish mineralogist Axel Cronstedt (1722-1765) and the name given as an abbreviated form of kopparnickel, Swedish false copper, since the ore in which it is found resembles copper but yields none.

Cobalt has very much in common with nickel, its neighbour in the periodic table. First of all, nickel is also of "devilish" origin. Its name derives from the German "Kupfernickel" ("copper devil") and belongs to the mineral described in 1694 by the Swedish mineralogist U. Hierne, who mistook it for copper ore. When repeated attempts to smelt copper from it failed, the metallurgists decided that it must have been nick, the evil spirit of the mountains, at his tricks.

People came to know nickel ages ago. Back in the 3rd century B.C. the Chinese made an alloy of copper, nickel and zinc. In the Central Asian state of Bactria, coins were made from this alloy. One of them is now in the British Museum in London.

Confusion about the composition of Kupfernickel remained even after the mineral had been described. In 1726 the German chemist I. Link studied the mineral and established that its dissolution in nitric acid yields a green colour. He concluded that the mineral was most probably a cobalt ore with admixtures of copper. When Swedish miners found a reddish mineral which, being added to glass, did not produce a blue colour, they named it "cobold that had lost his soul". It was also one of the nickel minerals.

That was how matters stood up to 1751. That year the Swedish mineralogist and chemist A. Cronstedt took an interest in the mineral found in a cobalt mine. In one of his experiments, he immersed a small piece of iron into an acid solution of this ore. Had copper been present in the solution, it would have been deposited on iron in a free state. To his great surprise nothing of the kind happened. The solution did not contain copper. This contradicted the existing beliefs about this ore. Cronstedt began a thorough investigation of the green crystals dispersed in the ore. After a great number of experiments, he isolated a metal from kupfernickel which did not resemble copper at all. Cronstedt described this metal as solid and brittle, weakly attracted by a magnet, transforming into a black powder when heated and yielding a wonderful green colour upon dissolution. Cronstedt concluded that, since the metal was contained in kupfernickel, the name could be retained and shortened to nickel. At present it is known that kupfernickel is nickel arsenide.

Many chemists in Europe recognized that a new element had been discovered. But some scientists held that nickel was a mixture of cobalt, iron, arsenic and copper. All doubts were removed in 1775 by T. Bergman who showed that mixtures of these elements taken in any proportions did not possess the properties of nickel.

MANGANESE

Manganese - hard, brittle, grey-white metallic element, symbol Mn, atomic number 25, relative atomic mass 54.9380. It resembles iron (and rusts), but it is not magnetic and is softer. It is used chiefly in making steel alloys, also alloys with aluminium and copper. It is used in fertilizers, paints and industrial chemicals. It is a necessary trace element in human nutrition. The name is old, deriving from the French and Italian forms of Latin for magnesia (Mgo), the white tasteless powder used as an antacid from ancient times.

Manganese compounds and in particular, its oxide - pyrolusite (MnO_2) – have been known from ancient times and used for making glass and pottery. In 1540 V. Biringuccio, an Italian metallurgist, wrote that pyrolusite was brown, did not melt and gave a violet colour to glass and ceramic when added to them. Another characteristic of pyrolusite was observed - its ability to make clear yellow and green glasses.

The Austrian Scientist I. Kaim seems to be the first to have obtained a small amount of metallic manganese in 1770. He heated a mixture consisting of one part of pyrolusite powder and two parts of black flux (i.e. a mixture of coal with K_2CO_3) and obtained bluish-white brittle crystals. Apparently, it was contaminated manganese but the scientist only concluded that the metal obtained was iron-free and did not complete his studies.

The subsequent story of manganese is associated with T. Bergman who by that time had already confirmed the discovery of nickel. He characterized pyrolusite in the following way: the mineral called "black magnesium" is a new earth; it should not be confused either with roasted lime or with "magnesium alba" (i.e. magnesium oxide). However, T. Bergman failed to separate the metal from pyrolusite, in contrast to I. Kaim.

C. Scheele was the third chemist who tried to separate a new element from this mineral. In 1774 he submitted his paper "On Manganese and Its Properties" to the Stockholm Academy of Sciences; in it he summed up the three years of studies of pyrolusite. In this extremely informative paper, he reported the discovery of two metals (barium and manganese) and described two gaseous elements (later identified as chlorine and oxygen). Scheele established that manganese oxide differed from all earth known at the time.

There are two significant dates in the history of manganese: May 16 and June 27, 1774. On May 16 Scheele sent I. Gahn, his friend and compatriot, a sample of purified pyrolusite and asked him to decompose it. Gahn placed a mixture of pyrolusite powder, oil and ground coal into a coal crucible and calcinated it for an hour. On the bottom of the crucible he found a regulus of the metal whose weight was only one third that of the initial pyrolusite. On June 27, having received from Gahn the sample of the new metal, Scheele wrote to his colleague that he considered the regulus obtained from pyrolusite a new semi-metal different from all other semi-metals and closely resembling iron. The term "Semi-metal" was retained in chemistry and metallurgy for some time. Thus, Gahn succeeded in separating metallic manganese. It may be said that he completed the discovery of this element, although manganese obtained by him had a high carbon content (pure metal was obtained later).

In 1785, the German chemist J. Ilsemann obtained metallic manganese independently of Gahn and Scheele by heating a mixture of pyrolusite, fluorspar, lime and coal powder. The product was intensely calcinated. The resulting manganese was, moreover, even less pure. At first the metal was named in Latin "manganesium" which derived from the old name of pyrolusite "Lapis manganensis". When in 1808 magnesium was obtained, to avoid confusion the Latin name of manganese was changed for "manganum".

BARIUM

Barium (Greek barytes "heavy") - soft, silver-white, metallic element, symbol Ba, atomic number 56, relative atomic mass 137.33. It is one of the alkaline-earth metals, found in nature as barium carbonate and barium sulphate. As sulphate it is used in medicine: taken as a suspension (a "barium meal"), its progress is followed by using X-rays to reveal abnormalities of the alimentary canal. Barium is also used in alloys, pigments and safety matches and with strontium, forms the emissive surface in cathode-ray tubes. It was first discovered in barytes or heavy spar.

Barium, as well as his analogues in the second group of the periodic table, are not encountered in nature in the native state. Sulphates and carbonates are the most typical barium minerals. One of barium minerals attracted attention of alchemists back in the early 17th century (in 1602, to be exact).

In that year V. Casciaralo, a shoemaker from Bologna, noted that heavy spar (barium sulphate), heated with coal and drying oil and then cooled to room temperature began to emit a reddish glow. The mineral, named Bologna stone, Bologna phosphorus, sunstone and so on, was barium sulphide BaS. The unusual luminescence was immediately interpreted in many different ways. For instance, the French chemist N. Lemery wrote in his "Chemistry Course" that the ability of Bologna stone to luminesce in the dark is due to the presence of sulphur. Another mineral displaying this property is Bolduin phosphorus (anhydrous calcium nitrate).

For a long time (up to 1774) heavy spar was confused with limestone; they were believed to be two varieties of the same compound. In 1774 Scheele, studying pyrolusite together with Gahn, discovered a new compound which gave a white precipitate under the action of sulphuric acid. Scheele established that heavy spar contained an unknown earth which was named "baryta" one.

By the last quarter of the 18th century, barium oxide was known rather well; it was suggested that it contained an unknown metal. This belief was supported by A. Lavoisier in his "Textbook of Chemistry". In "The Table of Simple Bodies" barite is considered a simple substance. However, only in 1808 did H. Davy succeed in preparing the new metal by the method which he had used for obtaining calcium.

The name of barium originates from the Greek baros (heavy) since barium oxide and its minerals were primarily characterized by their great mass.

MOLYBDENUM

Molybdenum (Greek malybdos "lead") - heavy, hard, lustrous, silver-white, metallic element, symbol Mo, atomic number 42, relative atomic mass 95.74. The chief ore is the mineral molybdenite. The element is highly resistant to heat and conducts electricity easily. It is used in alloys, often to harden steels. It is a necessary trace element in human nutrition. It was named in 1781 by Swedish chemist Karl Scheele, after its isolation by P. J. Hjelm (1746-1813), for its resemblance to lead ore.

It has a melting point of 2,620°C and is not found in the free-state. As an aid to lubrication, molybdenum disulphide (MoS_2) greatly reduces surface friction between ferrous metals. Producing countries include Canada, USA and Norway.

The molybdenum story is not rich in events. It is even trivial. Only one detail is of interest: this rare element was discovered very early, namely, in 1778, when the chemical analysis was just coming of age. Molybdenum was first separated in the form of oxide. The name "molybdenum" had appeared long before the new element was discovered. It originates from the Greek names molybdena for a lead mineral (lead glance) and molybdos for "lead", the two resembling each other. There was another mineral which also resembled these two very much; later it became known as molybdenite (molybdenum sulphide).

In 1754, the Swedish mineralogist A. Cronstedt differentiated these minerals, saying that molybdenite possessed some peculiar properties. But proof of that was required. By a lucky coincidence, the report of the molybdenite study fell into Scheele's hands. In 1778 he performed an analysis of molybdenite. The treatment of molybdenite with strong nitric acid resulted in the formation of a bulky white mass which Scheele described as a peculiar white earth. At the same time, nitric acid had no effect on graphite. Thus, the difference between graphite and molybdenite became evident. Scheele named the white earth "molybdic acid" since it had acid properties. Having calcinated molybdic acid, the Swedish chemist obtained molybdenum oxide, i.e. an oxide of a new metal. This is what Scheele believed and his belief was shared by his compatriot T. Bergman.

After that it was important to extract the metal from the molybdic earth. To do that Scheele planned to calcinate the earth with coal. But for some reason he could not perform the reaction himself and asked his friend P. Hjelm to do it. In 1790 Hjelm complied with the request. However, molybdenum obtained by him was contaminated with carbon and molybdenum carbide. The credit for preparing pure molybdenum (by reduction of the oxide with hydrogen) went to J. Berzelius (1817).

TUNGSTEN

Tungsten (Swedish tungsten "heavy stone") - hard, heavy, grey-white, metallic element, symbol W (from German wolfram), atomic number 74, relative atomic mass 183.85. It occurs in the minerals wolframite, scheelite and hubertite. It has the highest melting point of any metal (6,170°F/3410°C) and is added to steel to make it harder, stronger and more elastic; its other uses include high-speed cutting tools, electrical elements and thermionic couplings. Its salts are used in paint and tanning industries.

Tungsten was first recognized in 1781 by Swedish chemist Karl Scheele in the ore scheelite. It was isolated in 1783 by Spanish chemists Fausto D' Elhuyar (1755-1833) and his brother Juan Jose (1754-1796).

Although tungsten is also a rare element, it was discovered (in the form of its oxide) as early as the last quarter of the 18th century. To some extent, it was a matter of chance but progress in analytical chemistry also contributed to the discovery of tungsten.

The name "tungsten" appeared much earlier. In German it means "wolf's froth". The point is that in smelting of tin from some ores, a part of smelted metal was irretrievably lost. Medieval miners believed that tin was "devoured" by the mineral that was contained in the ore like a sheep is devoured by a wolf. This mineral was named tungsten or wolframite.

As time passed, tungsten attracted ever increasing attention of the scientists. In 1761 the German mineralogist I. Lemann analysed wolframite but did not find any new components in it. His compatriot P. Wolf for his part said that wolframite contained "something". Another strange mineral, "tungsten" or "heavy stone", was also known. It was found in 1751 by A. Cronstedt. In 1781 this mineral attracted the attention of C. Scheele who treated tungsten (calcium wolframate) with nitric acid and obtained a white substance resembling molybdic acid. An analyst par excellence, Scheele showed the difference between the two acids and consequently, he is considered to be the discoverer of tungsten.

At the same time T. Bergman, Scheele's compatriot, was also at the threshold of discovery. In his opinion, tungsten, due to its high density, could contain baryta earth. Studying the mineral, the scientist found a white substance in it which he called tungstic acid. But after that Bergman followed a wrong path, believing that both tungstic and molybdic acids were arsenic derivatives; however, he did not check this assumption. In 1783 two Spanish chemists, the F. and H.D' Egluar (brothers), separated a white acid from wolframite which proved to be similar to tungstic acid. Like Bergman and Scheele, the Spanish chemists succeeded in extracting metallic tungsten.

TELLURIUM

Tellurium (Latin Tellus "Earth") - silver-white, semi-metallic (metalloid) element, symbol Te, atomic number 52, relative atomic mass 127.60. Chemically, it is similar to sulphur and selenium and it is considered as one of the sulphur group. It occurs naturally in telluride minerals and is used in colouring glass blue-brown, in the electrolytic refining of zinc, in electronics and as a catalyst in refining petroleum.

It was discovered by Austrian mineralogist Franz Muller (1740-1825) in 1782, and named 1798 by German chemist Martin Klaproth.

Its strength and hardness are greatly increased by addition of 0.1% lead; in this form it is used for pipes and cable sheaths.

In the second half of the 18th century, a strange bluish-white ore was discovered in Austria or, to be more exact, in the part of it that was called Siebengebirge (Seven Mountains). It was strange because there was no common opinion about its composition. The debates mainly revolved around the question whether it contained gold or not. Its names were also unusual: paradoxical gold, white gold and finally, problematic gold. Some scientists believed that there was no problem at all and the ore, most likely, contained antimony or bismuth, or both. In 1782 the mining engineer I. Muller (later Baron Von Reichenstein) subjected the ore to a thorough chemical analysis and extracted metal reguluses from it which, as it seemed to him, closely resembled antimony. But in the following year he decided that in spite of the resemblance, he was dealing with a new, previously unknown metal. Not relying upon his own opinion, the scientist consulted T. Bergman. But the sample of the ore sent to Bergman was too small to come to a definite conclusion. It was only possible to establish that Muller's metal was not antimony and that was the end of the matter. During the next fifteen years, nobody recalled the discovery of the Austrian mining engineer. Tellurium's real birth was still ahead.

Its second birth was promoted by the German chemist M. Klaproth. At the Berlin Academy of Sciences session on January 25, 1798, he reported about the gold-bearing ore from "Seven Mountains". Klaproth repeated what Muller had done in his time. But if the latter was in doubt there was no doubt for M. Klaproth. He named the new element "tellurium" (from the Latin tellus for "Earth"). Although Klaproth had received the reo sample from Muller, he did not want to share the glory of the discoverer of tellurium with him; we, for our part, think that the role of the German chemist was no less important. At any rate he revived the forgotten element.

There is reason to believe that a third person was also involved in the discovery of tellurium. He was P. Kiteibel, a Professor of the Pest University in Hungary, a chemist and botanist. In 1789, he received a mineral which was assumed to be molybdenite containing silver from a colleague. P. Kiteibel extracted a new element from it. Then he established that the same element was present in problematic gold. Thus, P. Kiteibel discovered tellurium independently of other scientists. It is a pity that he did not publish at once his findings but instead sent a description of his investigation to some of his colleagues and in particular, to the Viennese mineralogist F. Estner. M. Klaproth learned about Kiteibel's results through F. Estner and spoke favourably of them without actually corroborating them. I. Muller wrote to M. Klaproth several years later and the latter found time to reproduce the results of his correspondent. After that, Klaproth considered him to be the only author of the discovery and he underlined this in his report.

For a long time tellurium was regarded as a metal. In 1832 Berzelius showed its great similarity with selenium and sulphur and tellurium was once and forever classified as a non-metal.

STRONTIUM

Strontium - soft, ductile, pale-yellow, metallic element, symbol Sr, atomic number 38, relative atomic mass 87.62. It is one of the alkaline-earth metals, widely distributed in small quantities only as a sulphate or carbonate, strontium salts burn with a red flame and are used in fireworks and signal flares.

The radioactive isotopes Sr-89 and Sr-90 (half-life 25 years) are some of the most dangerous products of the nuclear industry; they are fission products in nuclear explosions and in the reactors of nuclear power plants. Strontium is chemically similar to calcium and deposits in bones and other tissues, where the radioactivity is damaging. The element was named in 1808 by English chemist Humphry Davy, who isolated it by electrolysis, after Strontian, a mining location in Scotland where it was first found.

In 1787 a new mineral, strontianite, was found in a lead mine near the village of Strontian in Scotland. Some mineralogists classified it as a variety of fluorite (CaF_2). The majority of scientists, however, believed that strontianite was a variety of witherite (barium mineral $BaCO_3$).

In 1790 the Scottish physician A. Crawford, thoroughly studied the mineral and came to the conclusion that the salt obtained by the action of hydrochloric acid on strontianite differed from barium chloride. It dissolved in water more readily and its crystals were of different shape. Crawford decided that strontianite contained a previously unknown earth.

At the end of 1791, the Scottish chemist T. Hope concerned himself with studying strontianite and established the difference between witherite and strontianite. Hope also noted that the strontium earth reacted with water more vigorously than quicklime; it dissolved in water much more readily than barium oxide and all strontium compounds turned the flame red. T. Hope proved that the new earth could not be a mixture of calcium and barium earths. Lavoisier suggested that the new earth was of metallic nature but only H. Davy succeeded in proving it in 1808.

The history of the discovery of strontium would be incomplete if we did not mention another scientist to whom, undoubtedly, a great deal of credit for studying strontianite should be given. He was the Russian chemist T.E. Lovits who concluded, independently of other scientists, that strontianite contained an unknown element. Lovits was the first to discover strontium in heavy spar. The method of preparing metallic strontium suggested by H. Davy did not yield a sufficiently pure product. It was only in 1924 that P. Danner (USA) obtained pure strontium by reducing its oxide with metallic aluminium or magnesium.

ZIRCONIUM

Zirconium (Germanic Zircon, from Persian Zargun "golden") - lustrous, greyish-white, strong, ductile, metallic element, symbol Zr, atomic number 40, relative atomic mass 91.22. It occurs in nature as the mineral zircon (Zirconium silicate), from which it is obtained commercially. It is used in some ceramics, alloys for wire and filaments, steel manufacture, and nuclear reactors, where its low neutron absorption is advantageous.

It was isolated in 1824 by Swedish chemist Jons Berzelius. The name was proposed by English chemist Humphry Davy in 1808.

Zirconium oxide closely resembles aluminium oxide or alumina. For a long time the latter effectively concealed the presence of the former. Nobody suspected an unknown element in Zirconium minerals known as early as the Middle Ages. Thus, zirconium, one of the most abundant metals on Earth (0.02%) remained "invisible" up to the end of the 18th century. Today the mineral zircon is the main source of Zirconium; it occurs in two varieties: hyacinth and jargoon. Already in old times hyacinth was known as a precious stone owing to its beautiful colours ranging from yellow-brown to smoky green.

It was believed that the composition of hyacinth was similar to that of ruby and topaz.

Zircon was analysed more than once and every time erroneously. In 1787 the German chemist J. Wiegleb, when analysing Ceylon Zircon, found only silicon dioxide and small admixtures of lime, magnesia and iron. Earlier such a skilled chemist as T. Bergman had established that Ceylon hyacinth contained 25% silicon dioxide, 40% aluminium oxide, 13% iron oxide, and 20% lime. The element known subsequently as zirconium was safely "hidden" in aluminium oxide.

This natural camouflage was revealed in 1789 by M. Klaproth. He heated zircon powder (a sample similar to that used by T. Bergman) with alkali in a silver crucible. The alloy was then dissolved in sulphuric acid and from the solution M. Klaproth separated a new earth which he named zirconium. His analytical results demonstrated 25 percent silica, 0.5 percent iron oxide, 70 percent zirconium earth. As we see, there is nothing in common with Bergman's results. In the same year Guyton de Morveau, separating zirconium from hyacinth found in France, confirmed Klaproth's results.

Preparing metallic zirconium turned out to be not so simple. In 1808 H. Davy tried in vain to decompose zirconium earth with electric current. It was not before 1824 that Berzelius obtained contaminated zirconium by heating a dry mixture of potassium, potassium fluoride and zirconium in a platinum crucible. Zirconium received its name from the mineral.

URANIUM

Uranium - hard, lustrous, silver-white, malleable and ductile, radioactive, metallic element of the actinide series, symbol U, atomic number 92, relative atomic mass 238.029. It is the most abundant radioactive element in the Earth's crust, its decay giving rise to essentially all radioactive elements in nature; its final decay product is the stable element lead. Uranium combines readily with most elements to form compounds that are extremely poisonous. The chief ore is pitchblende, in which the element was discovered by German chemist Martin

Klaproth in 1789; he named it after the planet Uranus, which had been discovered in 1781.

Small amounts of certain compounds containing uranium have been used in the ceramics industry to make orange-yellow glazes and as mordants in dyeing; however, this practice was discontinued when the dangerous effects of radiation became known.

Uranium is one of the three fissile elements (the others are thorium and plutonium). It was long considered to be the element with the highest atomic number to occur in nature. The isotopes U-238 and U-235 have been used to help determine the age of the Earth.

Uranium-238, which comprises about 99% of all naturally occurring uranium, has a half-life of 4.51×10^9 years. Because of its abundance, it is the isotope from which fissile plutonium is produced in breeder nuclear reactors. The fissile isotope U-235 has a half-life of 7.13×10^8 years and comprises about 0.7% of naturally occurring uranium; it is used directly as a fuel for nuclear reactors and in the manufacture of nuclear weapons.

Many countries mine uranium; large deposits are found in Canada, USA, Australia and South Africa.

There is hardly another chemical element that from near oblivion sprang to instant fame. This is uranium occupying box no. 92 in the periodic table. Discovered in 1789, uranium did not interest chemists for a long time and even its atomic mass was determined incorrectly. Its practical use was confined to making coloured glass. But in 1906 in the eighth edition of Principles of Chemistry, Mendeleev appealed to those who were searching for new subjects of investigation to pay close attention to uranium compounds. The reason Mendeleev gave was that two most important events at the end of the 19th century science were related to uranium: the discovery of helium and the discovery of radioactivity. And, finally, is it a mere chance that uranium is the last in the series of naturally occurring chemical elements, the heaviest of them?

Some scientists have referred to the ninety second element as element no. 1 of our century.

And yet, there was nothing extraordinary about the discovery of uranium about two hundred years ago. It was like many others during the emergence of analytical chemistry. There is no doubt about the name of the discoverer - M. Klaproth. True, the actual extraction of uranium is associated with the name of another scientist (we shall come back to it later).

Pitchblende had been known to man for ages. When chemical analysis was still in its infancy, pitchblende was considered to be an ore of zinc and iron. More accurate knowledge of its composition was to come later.

When a pitchblende sample fell into the hands of Klaproth, he dissolved a piece of the mineral in nitric acid and added potash to the solution. Yellow precipitate was formed which was soluble in the excess of potash. The precipitate was small greenish-yellow crystals in the form of hexagonal prisms. Gradually, the scientist made the conclusion that he had obtained a salt of a new element. Having prepared an oxide, the scientist tried to separate pure metal. And when a lustrous black powder was formed at the bottom of the crucible, the German scientist decided that the aim was attained. But Klaproth was mistaken. At the most he had obtained a mixture of oxide with a small amount of the metal. Indeed, chemists were yet to see how difficult it is to extract pure uranium.

Confident of success, M. Klaproth proposed the name "uranium" for the element discovered. The chemist wrote: "In old times only seven planets were known and thought to correspond to seven metals, and according to this tradition the new metal should rightfully be named after the planet which has been recently discovered." It was the planet Uranus discovered in 1781 by the English astronomer Herschel. After that it became fashionable to name newly discovered chemical elements after celestial bodies. Uranium had been included in the list of simple substances and made its way to chemical textbooks, but metallic uranium remained unobtainable for a long time to come. There were even scientists who were doubtful about the discovery by the German chemist. Six years after Klaproth's death (1817), J. Arfvedson, the pupil of Berzelius decided, perhaps, following his teachers' advice, to remove these doubts. He tried to reduce dark-green uranium oxide with hydrogen. Arfvedson believed that the initial material was the lower oxide (we know now that the Swedish scientist worked with U_3O_8). The reaction yielded a brown powder (UO_2). J. Arfvedson, however, thought that he extracted metallic uranium.

It was only in 1841 that the French Chemist E. Peligot succeeded with the aid of a new reduction method. He heated anhydrous uranium chloride mixed with metallic potassium in a closed platinum crucible and obtained a black metallic powder. Its properties noticeably differed from those which M. Klaproth used to ascribe to metallic uranium. Therefore, some historians of science associate the real discovery of uranium with the name of E. Peligot.

Ingots of the metal were produced by the French chemist A. Moissan who melted it in an electric furnace invented by him in which a very high temperature could be attained. The scientist produced the first ingot in May, 1896 and gave it to Becquerel. With the aid of the sample, A. Becquerel established that radioactivity is a property of the elemental uranium. This property attracted everybody's attention to uranium for the first time.

At one time uranium gave a lot of trouble to D.T. Mendeleev when the scientist was working on his periodic table. The atomic mass of uranium was considered to be 120 and therefore, uranium was placed in the third group as a heavy analogue of aluminium. But this allocation by no means agreed with the properties of uranium, Mendeleev concluded that the atomic weight had been determined incorrectly and proposed to increase it by 100 percent. This put uranium in Group VI under tungsten and made it the last element in the periodic table.

TITANIUM

Titanium - strong, lightweight, silver-grey, metallic element, symbol Ti, atomic number 22, relative atomic mass 47.90. The ninth most abundant element in the Earth's crust, its compounds occur in practically all igneous rocks and their sedimentary deposits. It is very strong and resistant to corrosion, so it is used in building high-speed aircrafts and spacecrafts; it is also widely used in making alloys, as it unites with almost every metal except copper and aluminium. Titanium oxide is used in high-grade white pigments.

The element was discovered 1791 by English mineralogist William Gregor (1761-1817) and was named by German chemist Martin Klaproth in 1796 after Titan, one of the giants of Greek mythology. It was not obtained in pure form until 1925.

W. Gregor was not a chemist. But sometimes this English Clergyman did chemical experiments since his hobby was mineralogy. From time to time W. Gregor studied the composition of various minerals and so succeeded in the work that afterwards J. Berzelius respected him as a prominent mineralogist.

One day, W. Gregor became interested in the composition of black sand whose deposit he found in the Menaccin valley on the territory of his parish. This black sand, resembling very much gunpowder, mixed with dingy-white sand of a different kind attracted W. Gregor's attention. Having separated specks of black sand, he analysed them; you will judge the carefulness of this analysis from the following figures: $40\frac{9}{16}$ percent (this

9/16 is especially impressive) is iron oxides; $3\frac{1}{2}$ percent is silica and 45 percent is accounted for by the compound described by Gregor as reddish brown lime. And $4\frac{15}{16}$ percent was lost during the analysis. In this list it is the reddish-brown lime that is of interest. It dissolved in sulphuric acid yielding a yellow solution. Under the action of zinc, tin, or iron, the solution turned purple. Gregor wrote an article, reporting his findings. Being very modest, he believed that his investigation was incomplete. He only set forth some facts the explanation of which was the privilege of more knowledgeable scientists.

His friend, mineralogist D. Hawkins, convinced Gregor that the black sand was a new unknown mineral. Such an opinion from a man who knew about mineralogy not less than Gregor, suggested to the latter that the black sand contained a new metallic substance. Gregor proposed to name it "menaccin" in the honour of the place where the sand had been found and the sand itself menaccite (or menacconite). Now this black sand is named ilmenite and has the formula $FeTiO_3$. All this goes to show that titanium was discovered in 1791 by W. Gregor.

But many historians of science believe that M. Klaproth was the discoverer of titanium although the merit of Gregor's work is unquestionable. But the English clergyman was too unambitious. Klaproth chose another way. Ofcourse, he read Gregor's report but did not immediately grasp its meaning. In 1795 Klaproth succeeded in separating an oxide of the new element from the mineral brought from Hungary. Now this mineral is known as rutile (TiO_2). The oxide separated by Klaproth and the menaccin earth found by Gregor turned out to be very much alike. Soon Klaproth established that he and Gregor had discovered the same element.

The German scientist named the element "titanium" from mythological "Titans"-- the sons of Ge (the goddess of Earth). Pure metallic titanium was obtained only in 1910.

CHROMIUM

Chromium (Greek chromos "Colour") - hard, brittle, grey-white, metallic element, symbol Cr, atomic number 24, relative atomic mass 51.996. It takes a high polish, has a high melting point, and is very resistant to corrosion. It is used in chromium electroplating, in the manufacture of stainless steel and other alloys and as a catalyst. Its compounds are used for tanning leather and for alums. In human nutrition, it is a vital trace element. In nature, it occurs chiefly as chrome iron ore or chromite ($FeCr_2O_4$). Kazakhstan, Zimbabwe and Brazil are sources.

The element was named in 1797 by the French chemist Louis Vauquelin (1763-1829) after its brightly coloured compounds.

Siberia may be said to be the birthplace of chromium as we shall see later; in the 18th century the mineral crocoite, known at the time as red lead ore, was found there. Some other chromium ores had been known much earlier. And this is not surprising since chromium is one of the abundant elements (0.02 percent of the total mass of the earth's crust). But it is not easy to separate chromium even in the form of oxide and for the time being this task was beyond the power of chemists. Although chromium compounds have different colours, this peculiar fact did not attract the attention of scientists to chromium minerals.

The only exception was crocoite. For the first time it was analysed in 1766 by the German chemist I. Lehmann who lived at the time in St. Petersburg. Treating the mineral with hydrochloric acid, the chemist obtained a beautiful emerald-coloured solution. But his conclusion was erroneous: crocoite contained lead contaminated with impurities. These impurities could only be chromium since crocoite is lead chromate PbC_rO_4. I. Lehmann was not destined to establish the composition of the mineral.

For the second time crocoite became the object of study in 1770 when P.S. Pallas, a St. Petersburg academician, was describing the Berezov mines in the Urals: "This lead ore comes in different colours but more often looks like cinnabar. The crystals of this heavy mineral shaped as irregular pyramids are imbedded in quartz like little rubies."

P.S. Pallas was a traveller, geographer and mineralogist, and not a chemist. But it was he who introduced crocoite to laboratories in Western Europe. A sample of the mineral fell into the hands of the well-known chemist L. Vauquelin.

Three decades passed since I. Lehmann had studied crocoite. During this time the scientists repeatedly tried to determine its composition but failed to find any new elements in it. The results obtained were very contradictory. For instance, there was an analyst who reported that lead ore contained molybdic acid, nickel, cobalt, iron and copper. In his first experiments L. Vauquelin also made mistakes and found lead dioxide, iron and alumina in crocoite.

In 1797 the French chemist decided to study crocoite more thoroughly. Step by step Vauquelin refuted the results of all the previous analyses and at last drew a conclusion that crocoite contained a new metal with properties quite different from those of other metals.

L. Vauquelin boiled powdered crocoite with potassium carbonate. The product was lead carbonate and a yellow solution which contained, in the scientist's opinion, a potassium salt of an unknown acid. The solution acquired bright and diverse colours when various reagents were added: mercuric salts yielded a red sediment, lead salts gave a yellow sediment, tin chloride turned the solution green. All these results convinced Vauquelin that he was dealing with a new element. Its separation in the form of oxide was rather simple after that.

Many years later, D.I. Mendeleev wrote in his Principles of Chemistry that the uralian red chromium ore, or chromium-lead salt, had given Vauquelin the means to discover chromium. Vauquelin derived this name from the Greek chroma meaning "colour" because of the bright colouring of its compounds. For the sake of justice, we should note that the name "Chromium" for the new element was proposed by Vauquelin's compatriots A. Fourcroy and R. Hauy. Independently of Vauquelin and almost simultaneously with him, the presence of a new metal in crocoite was established by M. Klaproth who, however, did not prove it as clearly as his French colleague.

Numerous attempts to obtain pure chromium were unsuccessful. L. Vauquelin himself tried to prepare it but most likely it was chromium carbide that he had obtained.

BERYLLIUM

Beryllium - hard, light-weight, silver-white, metallic element, symbol Be, atomic number 4, relative atomic mass 9.012. It is one of the alkaline-earth metals, with chemical properties similar to those of magnesium; in nature it is found only in combination with other elements. It is used to make sturdy, light alloys and to control the speed of neutrons in nuclear reactors. Beryllium oxide was discovered in 1798 by French chemist Louis-Nicolas Vauquelin (1763-1829), but the element was not isolated until 1828, by German chemist Friedrich Wohler and Antoine-Alexandre-Brutus Bussy independently. The name comes from Latin beryllus.

In 1992, large amounts of beryllium were unexpectedly discovered in six old stars in the Milky Way.

Academician A.E. Fersman, the outstanding Soviet geo-chemist, called beryllium one of the most remarkable elements having tremendous theoretical and practical importance. However, beryllium is not outstanding in any one of its qualities; it has typical properties of metals. What is really remarkable, is the extremely fortunate combination (as if purposely invented by nature) of different properties. Beryllium clearly illustrates how the history of a chemical element is affected by its properties. As regards its chemical behaviour, beryllium has much more in common with aluminium (its diagonally neighbouring element in the periodic table) than with magnesium, its direct analogue in the same group. That is why aluminium was masking the presence of beryllium (as well as of zirconium) in natural minerals for such a long time.

Because of a pronounced amphoteric nature of beryllium, all attempts to obtain beryllium compounds in a sufficiently pure form were unsuccessful for a long time. As a result, many properties of the element and especially its valence and atomic mass were determined incorrectly. Consequently, the place of beryllium in the periodic table was not definitely found for a very long time. Only after it had been firmly established that beryllium is bivalent, that the formula of its oxide is BeO, and atomic mass is 9.01, was it once and for all placed in the upper box of the second group. A great contribution to that was made by the Russian Scientist I.V. Avdeev.

The history of beryllium minerals goes far back into the past when such precious stones as beryls and emeralds were already known.

One of the first scientists to begin the study of beryls in 1779 was F. Achard, Professor of Chemistry at the Berline Academy of Sciences. Before that time he had become famous for developing an industrial method of making sugar from sugar beet. The German chemist performed six analyses. His results recalculated in modern terms show that beryls contain 21.7% silicon oxide, 60.05% aluminium oxide, 5.02% iron oxide, and 8.3% calcium oxide. The total was only 95.07% (five percent was missing!) but F. Achard had no comment on this.

Similar figures were obtained in 1785 by J. Bindheim: in his case the "calculations" yielded the sum of the components of 101 percent. So, nothing particular was found about beryls.

In 1797 M. Klaproth, who by that time had already discovered uranium, titanium, and zirconium proving himself an outstanding analyst, received from the Russian diplomat and author D. Golitsyn samples of Peruvian emeralds and analysed them. But M. Klaproth did not wind up with 100 percent either (66.25% silica, 31.25% alumina, 0.5% iron oxide, total 98%). The scientist did not know where 2 percent had disappeared and did not try to explain. So he was not fated to add the discovery of the fourth element to his record.

At the same time, in France, another analyst L. Vauquelin, no less skillful than M. Klaproth, was at work. Beginning with 1793 he continued to study beryls and emeralds. But Vauquelin found nothing except ordinary components (silica, alumina, lime, iron oxide). Later Vauquelin recalled how difficult it had been to recognize a new substance when its properties were so similar to those of already known ones. The scientist meant a close similarity between oxides of aluminium and unknown beryllium.

Anticipating the events a little, we shall call Vauquelin the real discoverer of beryllium. The logic of discovery was not simple and it, undoubtedly, does justice to the scientist. He reasoned in the following way: beryl and emerald are very much alike as regards their composition and the shape of crystals. The crystal shape is absolutely the same but what about composition? Vauquelin's predecessors found the same components (alumina, silica, lime) in both minerals but their content varied.

After the first unsuccessful experiments, L. Vauquelin decided to see why the components' content varied so widely. Could it be that the minerals contained "Something" else which was either lost in the course of the reaction or, figuratively speaking, was "hiding behind the backs" of one of the components (for instance, alumina).

L. Vauquelin had a certain psychological advantage. In 1797 he discovered chromium, which imparts a greenish colour to emerald and is absent in beryl. Hence, the difference between beryl and emerald is an established fact. But not only chromium could be responsible for the difference. February 14, 1798 should be considered as the birthday of beryllium. On that day Vauquelin made a report to the Paris Academy of Sciences, "About Aquamarine, or Beryl and the Discovery of a New Earth in This Mineral." He told the audience how he had performed five analyses and how he had become more and more convinced of the existence of the new earth. The first results were as follows:

Beryl: 69 parts of silica, 21 parts of alumina, 8-9 parts of lime, and 1½ parts of iron oxide.

Emerald: 64 parts of silica, 29 parts of alumina, 2 parts of lime, 3-4 parts of chromium oxide and 1-2 parts of water.

Whether it was intuition or something else, but Vauquelin suspected that in both cases alumina contained an impurity. It resembled alumina very much and, therefore, it was rather difficult to detect it. The brilliant intuition of an analyst helped the scientist to discover that the impurity (the new earth), unlike alumina, did not form alum. Later he found other differences. But similarity prevailed over difference enabling beryllium to hide for so long behind aluminium. If beryllium earth is not alumina, L. Vauquelin thought, it is none of the known earths since it differs from them much more than alumina. The scientist proposed to name the new element "glucinium" (symbol GI) from the Greek glykys which means "Sweet". The present name "Beryllium" was proposed by M. Klaproth who justly noted that some compounds of other elements are also sweet.

As an interesting historical detail, we should like to mention that Vauquelin analysed Altaian beryls presented to him by French mineralogist and traveller E. Patren.

The discovery of L. Vauquelin was confirmed by I. Gmelin, the German chemist, a professor of chemistry in Gottingen. He analysed Siberian beryls from Nerchinsk and made the same conclusions as Vauquelin. Metallic beryllium was isolated in 1828 by F. Wohler and E. Bussy who treated beryllium chloride with potassium metal. It was thirty years after the discovery of beryllium.

NIOBIUM AND TANTALUM

Niobium - soft, grey-white, somewhat ductile and malleable, metallic element, symbol Nb, atomic number 41, relative atomic mass 92.906. It occurs in nature with tantalum, which it resembles in chemical properties. It is used in making stainless steel and other alloys for jet engines and rockets and for making superconductor magnets.

Niobium was discovered in 1801 by English chemist Charles Hatchett (1765-1847) who named it columbium (symbol Cb), a name that is still used in metallurgy. In 1844 it was renamed after Niobe by German Chemist Heinrich Rose (1795-1864) because of its similarity to tantalum (Niobe is the daughter of Tantalus in Greek Mythology)

Tantalum - hard, ductile, lustrous, grey-white, metallic element, symbol Ta, atomic number 73, relative atomic mass 180.948. It occurs with niobium in tantalite and other minerals. It can be drawn into wire with a very high melting point and great tenacity, useful for lamp filaments subject to vibration. It is also used in alloys, for corrosion-resistant laboratory apparatus and chemical equipment, as a catalyst in manufacturing synthetic rubber, in tools and instruments, and in rectifiers and capacitors.

It was discovered and named in 1802 by Swedish chemist Anders Ekeberg (1767-1813) after the mythological Greek character Tantalus.

The early histories of these elements are so intertwined that it is hardly worthwhile to consider them separately. Their

81

common history begins on November 26, 1801, when Ch. Hatchett made a report to a session of the Royal Society about the discovery of a new element. Communications of this type had long ceased to be a sensation. Hatchett's report "Analysis of a Mineral from North America Containing an Unknown Metal" was received quietly. True, Hatchett got his sample not from the New World but from a place much nearer - from the British Museum. The Museum's catalogue described the mineral as "a black ore sent to the Museum by Wintrop from Massachusetts".

At first Ch. Hatchett assumed that the object of his study was a variety of Siberian chromium ore and tried to isolate chromic acid from it. But things took a different turn. Now it is known that the mineral from Massachusetts contained a variety of metals and it was not easy to extract a new element from it. There was no chromium in the mineral and Hatchett concluded that the compound which he had separated was not chromic acid but an oxide of an unknown metal. In honour of its place of origin, the English Scientist named the mineral "columbite" (from Columbus and Columbia, the former name of America). The element was named "columbium". A year later, in 1802, an event took place which added a little zest to the trivial discovery of columbium. In December 1802 the Swedish chemist A. Ekeberg, who analysed some minerals found near the village of Itterbul, described the discovery of an oxide of a new metal. The white oxide mass did not dissolve even in a great excess of strong acids.

The futility of all attempts to dissolve the oxide prompted Ekeberg to name the new metal "tantalum" after the "torments of Tantalus" which means useless and futile work. The mineral was named "tantalite". A. Ekeberg was firmly convinced that he had discovered a new element and this conviction was shared by many scientists. The more surprising were the results of the English chemist V. Wollaston who announced in 1809 that he found no difference between columbium and tantalum and that the two were one and the same element. Oxides of these metals had similar densities and seemed to Wollaston to be rather similar in their chemical properties. His article was titled "On the identity of Columbium and Tantalum". This meant that A. Ekeberg had only rediscovered columbium, confirming the discovery made by Ch. Hatchett.

Berzelius held a different opinion. He supported the name "tantalum" given to the new element by Ekeberg and believed that the names of the English and Swedish chemists must stand together in history. In autumn 1814, Berzelius wrote in a private letter to the Scottish chemist Th. Thomson (the first advocate of Dalton's atomic theory) that he by no means wanted to belittle Hatchett's achievement but deemed it his duty to note that the properties of tantalum and its oxide had been almost unknown before Ekeberg's work. Berzelius thought that the columbic acid of Hatchett was a mixture of tantalum oxide and tungstic acid, but soon it became clear that there was no tungsten in columbite.

Three decades later one of Berzelius pupils, H. Rose, resolved the dispute once and for all. He proved that tantalum and columbite were not identical; hence, Hatchett and Ekeberg had discovered two different elements.

Rose analysed columbites and tantalites from different deposits. And every time he found that, along with tantalum, they contained another element whose properties were close to those of tantalum. Rose named the "stranger" "niobium" (Niobe was Tantalus' daughter). In the summer of 1845, the scientist studied the same mineral in which Hatchett had once detected columbium and isolated niobium oxide from it, which proved to be similar to Columbium oxide.

At last, the confusion was cleared. It had arisen because niobium and tantalum have very similar properties and are always present together both in columbites and tantalites.

As a matter of fact. Hatchett and Ekeberg discovered both elements simultaneously and could not detect any difference between them. In the mineral studied by Hatchett, niobium (columbium) undoubtedly predominated. Therefore, the most important event in the biography of both elements was the development of a method for separating niobium and tantalum. This was done in 1865 by the Swiss chemist J.C. Galissard de Marignac who found the difference in solubilities of potassium fluotantalate and fluoniobate in hydrofluoric acid. In the same year de Marignac correctly determined the atomic masses of niobium and tantalum for the first time. Many chemists tried to obtain them in a pure state but, as a rule, wound up with contaminated metals. Not before the beginning of the 20th century did W. Von Bolten of USA obtain niobium and tantalum of higher than 99 percent purity.

PLATINUM METALS

The history of platinum metals (ruthenium, rhodium, palladium, osmium, iridium and platinum) is full of false discoveries of chemical elements made in the studies of these metals and due to the great difficulties involved in studying natural ores containing platinum and accompanying metals. Platinum that mankind had come to know prior to the real discovery of this element contained different impurities. Among the platinum metals, platinum occupies the second place after palladium in terms of abundance. The content of platinum metals in the minerals may vary considerably from deposit to deposit. Therefore, there were many chance events in the history of platinum and its analogues and much is still unclear. The date of the discovery of platinum is rather vague. For a long time, it was not clear how many platinum metals really exist. In many cases the confusion arose because of similar properties of platinum metals. Four of them - palladium, rhodium, osmium, and iridium - were discovered in the early 19th century owing to the considerable progress in chemical analysis. However, it is quite possible that it was just a chance that prevented earlier discovery of platinum metals, at any rate of the sufficiently abundant palladium.

PLATINUM

Platinum (Spanish Platina "little silver") - heavy, soft, silver-white, malleable and ductile, metallic element, symbol Pt, atomic number 78, relative atomic mass 195.09. It is the first of a group of six metallic elements platinum, osmium, iridium, rhodium, ruthenium, and palladium) that possess similar traits, such as resistance to tarnish, corrosion, and attack by acid, and that often occur as free metals (native metals). They often occur in natural alloys with each other, the commonest of which is osmiridium. Both pure and as an alloy, platinum is used in dentistry, jewelry, and as a catalyst.

Platinum was the first to be discovered among the platinum metals. 1748 is considered to be its date of birth. But is it the real date?

Ancient Greeks and Romans mentioned "electrum", an alloy that some scientists identify with platinum. Others believe that "electrum" was the Egyptian alloy of gold with silver. Pliny the Elder described a white heavy compound found in the sands of Galicia and Portugal but it was, most likely, a tin ore. A box made of platinum was found in the tomb of Queen Shapenapit (the 7th century B.C.).

In 1557 the Italian Scientist G. Scaliger described a new white metal discovered in South America. It was the first definite mention of platinum. Another two centuries passed. The Paris Academy of Sciences sent an expedition to the Spanish colonies.

Among its participants was a young lieutenant Don Antonio de Ulloa. Having safely returned home, he wrote the book "Historical Report about the Trip to South America" which was published in Madrid in 1748. He wrote that in the region of Choko he had seen many gold-bearing mines but some of them had been abandoned because of a high content of platinum in the ore. A. Ulloa was the first to note that this metal had an extremely high melting point and that it was very difficult to extract it from the ores. Two years later the English chemists W. Watson and W. Brownrigg set out to study the new metal and gave the first scientific description of it. In November 1750, W. Watson reported the discovery of a new semi-metal called "platino-del-pinto" which had hitherto been unknown to mineralogists.

This work prompted further study of the new metal. In 1752 the Swiss chemist H. Scheffer published a detailed report about his investigation of platinum or white gold. After that a series of similar papers appeared. Two of them were particularly interesting. In 1772 C. Von Sickingen extensively studied the properties of platinum, looking into the possibility of alloying platinum with silver and gold, its solubility in aqua regia, and what is most important, he was the first to use the method of precipitating platinum from solutions with ammonium chloride. This reaction played a great role in studying the platinum metals. But the results obtained by C. Von Sickingen were not published until 1782.

The second round of studies is associated with the name of P. Chabanean. He was the first to pay attention to the fact that experiments with platinum from different deposits yielded contradictory results. With hindsight this has a very single explanation: Chabanean was working not with pure platinum but with a mixture of six elements - the platinum metals that had not yet been discovered. For instance, in the absence of osmium, platinum was non-volatile and did not ignite whereas the presence of osmium made the alloy volatile and combustible.

What is the exact date of platinum's discovery? The metal had to go a long way before it was given the right to its own title. 1750 seems to be a major landmark in the history of platinum: in that year it was studied and described in detail.

PALLADIUM

Palladium - lightweight, ductile and malleable, silver-white, metallic element, symbol Pd, atomic number 46, relative atomic mass 106.4.

It is one of the so-called platinum group of metals and is resistant to tarnish and corrosion. It often occurs in nature as a free metal in a natural alloy with platinum. Palladium is used as a catalyst, in alloys of gold (to make white gold) and silver, in electroplating, and in dentistry.

It was discovered in 1803 by British physicist William Wollaston (1766-1828) and named after the then recently discovered asteroid Pallas (found 1802).

Back in the late 17th century Brazilian miners frequently ran into a strange naturally occurring alloy. It had different names and was believed to contain gold and silver. It could be an alloy of palladium and gold. But the real discovery of the second of the platinum metals took place in 1803 owing to the work of the English chemist W. Wollaston. Studying crude (unpurified) platinum, he dissolved it in aqua regia, removed the excess of the acid, and added mercury cyanide to the solution. A yellow precipitate was formed. Heating the solution with sulphur and borax, he obtained bright metal balls. Wollaston named the new metal "palladium" (after the asteroid discovered a year earlier by the astronomer W. Olbers). Wollaston's success was largely owing to the fact that he had found a proper precipitating agent

87

for palladium, mercury cyanide, which does not precipitate other platinum metals.

The discovery of palladium received publicity in a rather peculiar way. In 1804 the young Irish chemist R. Chenevix put an advertisement in the Journal of Chemical Education about a "New metal for Sale" which was an alloy of platinum with mercury. W. Wollaston, naturally, was of a different opinion and defended his discovery. He published the article "On a New Metal Found in Crude Platinum" in which he underlined that the metal "for Sale" named palladium is contained in platinum ores although in small amounts.

Contemporary Scientists (and L. Vauquelin among them) valued highly the Wollaston's achievement, and more so since soon he discovered another platinum metal, rhodium. The fact that palladium was the first platinum metal to be extracted may be explained by its greatest abundance as compared with other platinum metals. In addition, it exists in nature in a native state as proved by Wollaston in 1809 and by A. Humboldt in 1825 (for Brazilian platinum ores which had been the only source of material prior to the discovery of Uralian platinum).

RHODIUM

Rhodium (Greek rhodon "rose") - hard, silver-white, metallic element, symbol Rh, atomic number 45, relative atomic mass 102.905. It is one of the so-called platinum group of metals and is resistant to tarnish, corrosion and acid. It occurs as a free metal in the natural alloy osmiridium and is used in jewelry, electroplating, and thermocouples.

Rhodium was discovered in 1803 by English chemist William Wollaston (1766-1828) and named in 1804 after the red colour of its salts in solution.

The discovery of palladium became the key to the discovery of rhodium at the turn of 1803, i.e. before the news about palladium was widely spread.

Crude platinum from South America was also a source of rhodium. It is, however, not known whether it was the same sample in which Wollaston had discovered palladium. Having dissolved a certain amount of crude platinum in aqua regia and neutralized the excess of the acid with alkali, Wollaston first added an ammonium salt to precipitate platinum as ammonium chloroplatinate. Mercury cyanide was added to the remaining solution (here the experience in separating palladium proved useful) and palladium cyanide precipitated. Then Wollaston removed the excess of mercury cyanide from the solution and evaporated it to dryness; a beautiful dark-red precipitate was formed which, in the scientist's opinion, was double chloride of sodium and of the new metal.

This salt decomposed readily upon heating in a hydrogen flow, as a result of which metallic powder was formed (after removal of sodium chloride). The scientist also obtained the new metal in the form of pellets. The name "rhodium" was given to the new element because of the red colour of its first salt to be produced (the Greek rodon means "a rose").

This element is the least abundant of the platinum metals. The only rhodium mineral known is rhodite, found in gold-bearing sands of Brazil and Colombia, whereas several minerals are known for each of the other platinum metals.

OSMIUM AND IRIDIUM

Osmium (Greek osme "Odour") - hard, heavy, bluish-white, metallic element, symbol Os, atomic number 76, relative atomic mass 190.2. It is the densest of the elements and is resistant to tarnish and corrosion. It occurs in platinum ores and as a free metal with iridium in a natural alloy called osmiridium, containing traces of platinum, ruthenium and rhodium. Its uses include pen points and light bulb filaments; like platinum, it is a useful catalyst.

Osmium was discovered in 1803 and named in 1804 by English chemist Smithson Tennant (1761-1815) after the irritating smell of one of its oxides.

Iridium (Latin iridis "rainbow") - hard, brittle, silver white, metallic element, symbol Ir, atomic number 77, relative atomic mass 192.2. It is twice as heavy as lead and is resistant to tarnish and corrosion. It is one of the so-called platinum group of metals; it occurs in platinum ores and as a free metal (native metal) with osmium in osmiridium, a natural alloy that includes platinum, ruthenium and rhodium.

It is alloyed with platinum for jewelry and used for watch bearings and in scientific instruments. It was named in 1804 by English chemist Smithson Tennant (1761-1815) for its iridescence in solution.

The discovery of four new elements with similar properties in one country (England) in the course of two years was unprecedented in the history of science. Another English chemist, S. Tennant, was studying platinum metals simultaneously with W. Wollaston, who discovered palladium and rhodium while the extraction of osmium and iridium is associated with the names of other scientists, although the greatest contribution was made by S. Tennant.

As compared with other platinum metals, osmium and iridium have some specific features to which they owe their names. "Osmium" derives from the Greek osme for "smell" since osmium oxide is volatile and has a peculiar smell. Iridium got its name from the variety of colouring of its salts (from the Greek iris for "rainbow"). A painter could have prepared an entire palette from iridium paints if they were not so expensive. These unusual properties promoted the discovery of these platinum metals.

S. Tennant, like W. Wollaston, dissolved crude platinum in aqua regia. At the bottom of the retort he discovered a black precipitate with metallic lustre. This phenomenon had been observed previously in experiments with platinum, but the precipitate was believed to be graphite. In summer 1803 Tennant suggested that the precipitate most likely contained a new metal. In autumn of the same year the French chemist H. Collet-Descoties also concluded that the precipitate contained a metal that precipitated from ammonium platinum salts and yielded red colour. In his turn, L. Vauquelin heated the black powder with alkali and obtained a volatile oxide. Vauquelin believed that it was an oxide of the metal mentioned by H. Descoties. Tennant's experiment set off a series of investigations. Tennant himself continued his research and in spring 1804, he reported to the British Royal Society that the powder contained two new metals which could be separated fairly easily. In 1805, he published the article "On Two Metals Found in the Black Powder Formed after Dissolution of Platinum". The names "osmium" and "iridium" were mentioned in the article for the first time.

The notorious black powder was, evidently, a natural alloy of osmium with iridium, the so-called osmiridium. Iridium is known to be chemically stable and in the compact form does not dissolve even in aqua regia. On the contrary, osmium is readily soluble in aqua regia; in general among platinum metals, osmium has the most atypical chemical properties. That is why iridium and osmium were relatively quickly and easily separated.

In 1817, the English chemist and mineralogist W. Brande justly noted in his lecture devoted to the discovery of platinum metals that if one tried to analyse the entire development of chemistry from the standpoint of contemporary analytical accuracy, the history of the discovery and separation of platinum metals would, probably, be the most striking one.

But had all of platinum metals been discovered? The question was posed again and again. Years passed but they brought nothing new, at any rate, no reliable answer. Only in 1844 was ruthenium, the last of the platinum metals, discovered;

ruthenium is as abundant in nature as platinum, which, with its greatest atomic mass, was the first to be discovered. Why it was so remains a mystery. It may have been pure chance since the study of platinum metals was extremely difficult and required great analytical skill and profound knowledge of chemistry.

RUTHENIUM

Ruthenium - hard, brittle, silver-white, metallic element, symbol Ru, atomic number 44, relative atomic mass 101.07. It is one of the so-called platinum group of metals: it occurs in platinum ores as a free metal and in the natural alloy osmiridium. It is used as a hardener in alloys and as a catalyst; its compounds are used as colouring agents in glass and ceramics.

It was discovered in 1827 and named in 1828 after its place of discovery, the Ural Mountains in Ruthenia (now part of the Ukraine). Pure ruthenium was not isolated until 1845.

Ruthenium was the first chemical element discovered by a Russian scientist. It was Karl Klaus. The discovery of this last of the platinum metals was made forty years after the discovery of iridium.

In 1828 G.V. Ozann, Professor of the Tartu University, studied the residue obtained after the dissolution of crude uralina platinum in aqua regia and found that it contained three new elements: pluranium, polinium, and ruthenium. But Berzelius, to whom Ozann had sent a letter about his findings, did not support the discovery. Because of this significant fact the study of this platinum residue was not renewed until 1841. Berzelius's prestige was so high that no chemist in the world would argue with him.

The second reason for such a late discovery of ruthenium is its great similarity to the other "brothers"' in the family. Prior to Klaus in Russia, this problem was studied by the Polish scientist A. Snyadetskii who also reported the discovery of a new element which he named "West" after the asteroid of the same name. But his discovery proved to be false.

K. Klaus began his research in 1840. The then Minister of Finance of Russia E.F. Kankrin, a competent and energetic person, rendered him great assistance; Klaus obtained 2 pounds of crude platinum residue and extracted a considerable amount of iridium, rhodium, osmium and palladium from it, apart from 10% platinum. In addition, Klaus separated a mixture of metals which, in his opinion, had to contain a new substance.

First of all, Klaus repeated Ozann's experiments. Then he continued the investigation according to his own plan. The results were striking. In 1844 he published a 188-page report with the following information: analytical results on the residue obtained after platinum dissolution in aqua regia; new methods of platinum metals separation; methods of studying lean residues; the discovery of a new metal-ruthenium; analytical results on lean residues and the simple methods of separating platinum ores and residues; new properties and compounds of the previously known metals of the platinum group. This was a real encyclopaedia on chemistry of platinum metals.

K. Klaus separated six grams of the new element from its double salt with potassium. He sent a report about it to Berzelius but the latter was sceptical again. Great courage was required from Klaus to contradict the old and eminent scientist. The Russian chemist proved the genuineness of his discovery and in 1845. J. Berzelius recognized the new element. A special committee was formed in Russia consisting of academicians H. Hess and Yu.F. Fritsshe to check the results obtained by Klaus. The committee confirmed the discovery and K. Klaus was awarded the Demidov's prize (1000 roubles).

The name of the element is derived from the Latin for Russia (Ruthenia). Klaus gave this name to the element moved by his patriotic feelings and trying to show that all work in this field had been done in Russia (G. Ozann, A. Snyadetskii, K. Klaus).

Klaus spent a total of 20 years studying platinum metals. He deserves the right to be called the founder of the Russian school of studies of platinum and platinum metals.

HALOGENS

Man did not become properly acquainted with halogens until the 19th century although fluorine and chlorine were discovered in the seventies of the 18th century. But the fact that chlorine is a chemical element was understood only about forty years after its discovery. Fluorine was "hiding" behind fluorine compounds for a whole century before it was, at last, obtained in a free state. But iodine and bromine were at once recognized as simple substances.

As we see, the fates of these elements, named halogens in 1811, were different in the history of science but they played a peculiar role, especially in chemistry.

All of them were produced by chemical analysis except ree fluorine which was prepared electrochemically.

FLUORINE

Fluorine - pale yellow, gaseous, non-metallic element, symbol F, atomic number 9, relative atomic mass 19. It is the first member of the halogen group of elements and is pungent, poisonous, and highly reactive, uniting directly with nearly all the elements. It occurs naturally as the minerals fluorite (CaF_2) and cryolite (Na_3AlF_6). Hydrogen fluoride is used in etching glass, and the freons, which all contain fluorine, are widely used as refrigerants.

Fluorine was discovered by the Swedish chemist Karl Scheele in 1771 and isolated by the French chemist Henri Moissan in 1886. Combined with uranium as UF_6, it is used in the separation of uranium isotopes.

The famous Soviet scientist A.E. Fersman called this chemical element "omnivorous". And indeed, there are very few substances, both natural and man-made, that can withstand unprecedented chemical aggressiveness of fluorine. The story of fluorine is an illustration of this property. Fluorine proved to be the last (chronologically) non-metal to be separated in a free state (apart from inert gases). One hundred years passed from the time of the forecasting of the existence of fluorine to the moment when scientists succeeded in obtaining it in a gaseous state. Chemists tried to prepare it over fifteen times but every time the attempts failed. And in several cases they even lost their lives.

At the same time a common natural compound of fluorine (fluorspar or fluorite, CaF_2) had been known from very remote times. This harmless mineral as known to any stone collector was mentioned in manuscripts as early as the 16th century. But when hydrofluoric acid was first prepared, fluorite assumed new significance. It is difficult to establish who was the first to prepare hydrofluoric acid. All that is known is that in 1670 the Nurnberg craftsman H. Schwanhard observed its corrosive action on glass. Schwanhard and many after him erroneously believed that etching of glass was caused by silicic acid, while it was hydrofluoric acid that destroyed glass.

A century passed before fluorspar fell into the hands of C. Scheele. He studied two varieties of fluorite - green and white. The scientist heated powdered samples with sulphuric acid and noticed that the inner surface of the glass retort became opaque while a white mass precipitated on the bottom of the retort. Scheele assumed that fluorite consisted of lime earth saturated with unknown acid. He added lime water to this acid and obtained artificial fluorspar similar to the natural mineral.

The year when hydrofluoric acid was separated (1771) is considered to be the date of the discovery of fluorine although this is hardly justified. The nature of the acid obtained by Scheele (named "Swedish acid" at the time) remained unclear. There was a controversy in the scientific world about Scheele's discovery but with every year it became increasingly evident that he was right.

Hydrofluoric acid entered the category of reliably classified chemical compounds and scientists gradually came to believe that it contained a new chemical element. This opinion was strengthened by A. Lavoisier who included the radical of hydrofluoric acid (radical fluorique) as a simple body into "The Table of Simple Bodies". But Lavoisier was also wrong: he thought that the acid contained oxygen. His mistake was, however, understandable since at that time chemists believed that oxygen was an indispensable constituent of all acids.

The purity of the acid prepared by Scheele's method left much to be desired. Not before 1809 did Gay Lussac and Thenard obtain a relatively pure hydrofluoric acid, heating fluorspar with sulphuric acid in a lead retort. Both scientists were severely poisoned during the experiments.

A year later an event of extreme importance took place in the pre-history of fluorine. Two scientists-the Englishman H. Davy and the Frenchman A. Ampere - independently "banished" oxygen from hydrofluoric acid. They strongly believed that the acid was a compound of hydrogen with an unknown element and that it is similar to hydrochloric acid HCl. It was the second decisive intervention of H. Davy in the fate of halogens (shortly before he had established the elemental nature of chlorine).

It is therefore clear why Davy was the first who attempted to obtain free fluorine. By the way, the name was proposed by Ampere who borrowed it from the Greek ftoros for "destructive". Ampere chose this name because of the hydrofluoric acid's aggressiveness (Chemists were still to see the fury of free fluorine). But Davy was in a more peaceable mood and suggested the name of "fluorine" by analogy with "Chlorine".

Having named the element, Davy, nevertheless, did not succeed in preparing free fluorine. For two years (1813 and 1814) the scientist was storming the impregnable fortress. Two methods were used by H. Davy: the electrochemical method, which had already given the world sodium, potassium, calcium and magnesium and the reactions of chlorine with fluorides. Electrolysis of hydrofluoric acid gave no results; the second method was also fruitless. Severe illness caused by work with fluorine-containing compounds forced Davy to stop the experiments although he was one of the first to determine the atomic mass of fluorine (19.06). Davy's unsuccessful experiments and his illness seemed to serve as a warning for other scientists and for almost 20 years nobody tried to obtain free fluorine. Only M. Faraday, Davy's famous pupil and assistant, whose contribution to science was no less important than that of his teacher, made an attempt in 1834 (after Davy's death) to solve the riddle of free fluorine. However, even electrolysis of dry melted fluorides proved to be futile.

The chain of failures grew longer. In 1836, the brothers Knox from Ireland set out to solve the problem. During five years they were performing dangerous experiments, without success. The brothers were severely poisoned in the process and R. Knox died. In 1846, the Belgian chemist P. Layette and then the French chemist D. Niklesse shared the dramatic fate of the Knox brothers. At last, in 1854-1856, E. Fremy, Professor of Ecole Polythechnique in Paris, seemed to succeed in preparing free fluorine. He electrolytically decomposed anhydrous melted CaF_2. Metallic calcium deposited on the cathode, while on the anode a gas was liberated which could be nothing but fluorine. However, to observe a chain of bubbles is not enough - they had to be collected. In this, however, Fremy failed. But, in our opinion, E. Fremy deserves the name of a co-discoverer of fluorine, at any rate, his right to it is no less than that of Scheele.

In 1869, the English chemist G. Gore obtained a small amount of free fluorine which at once reacted explosively with hydrogen. There were about ten other researchers who hoped to obtain free fluorine. History, of course, has their names but we shall not mention them here.

And at last the moment came when A. Moissan took resolutely in his hands the fate of fluorine. First of all, he analysed the errors of his predecessors and clearly realized that the attempts of Faraday. E. Fremy and G. Gore had failed because they could not subdue the "Fury" of fluorine which instantly reacted with the material of the apparatus. Moissan was also aware of the mistake of those investigators who tried to isolate fluorine by the action of chlorine on fluorides; chlorine had to be a weaker oxidizer than fluorine.

Moissan overcame the difficulty by using a U-shaped vessel. At first he used a platinum vessel but later decided that a copper one must be much more suitable since neither fluorine nor hydrogen fluoride reacted with copper fluoride being formed. Thus, a layer of copper fluoride prevented the vessel from destruction. Moissan filled the vessel with anhydrous hydrofluoric acid and added a small amount of potassium bifluoride to it for the solution to become electro-conductive. The vessel was immersed in a cooling mixture at -25°C. Platinum electrodes were inserted through CaF_2 plugs. Electrolysis liberated hydrogen on the cathode and fluorine on the anode; fluorine was collected in copper tubes.

On June 26, 1886, Moissan performed the first successful experiment, observing the flame produced by the reaction of fluorine with silicon. He sent a modest report to the Paris Academy of Sciences where he wrote that different hypotheses about the nature of the liberated gas were possible. The simplest of them was that fluorine is actually liberated, although the gas might also be hydrogen perfluoride or even a mixture of HF and ozone. The reactivity of this mixture is high enough to explain the strong action of the gas on crystalline silicic acid.

Since Moissan was not a member of the Academy, his report was read by A. Debray and a special committee was organized consisting of A. Debray, E.Fremy, and "the Elder" of the French chemists M. Berthelot. On the first day, Moissan's attempt to prepare free fluorine failed but on the following day he succeeded and the committee witnessed his success. Thus, another date appeared in the biography of fluorine and, maybe, the most important one - the date of its preparation in a free state (1886). In 1887 Moissan obtained liquid fluorine.

CHLORINE

Chlorine (Greek Chloros "green") - greenish-yellow, gaseous, non-metallic element with a pungent odour, symbol Cl, atomic number 17, relative atomic mass 35.453. It is a member of the halogen group and is widely distributed, in combination with alkali metals, as chlorates or chlorides.

In nature, it is always found in the combined form, as in hydrochloric acid, produced in the mammalian stomach for digestion. Chlorine is obtained commercially by the electrolysis of concentrated brine and is an important bleaching agent and germicide, used for both drinking and swimming-pool water. As an oxidizing agent, it finds many applications in organic chemistry. The pure gas (Cl_2) is a poison and was used in gas warfare in World War 1, where its release seared the membranes of the nose, throat and lungs, producing pneumonia. Chlorine is a component of chlorofluorocarbons (CFCs) and is partially responsible for the depletion of the ozone layer; it is released from the CFC molecule by the action of ultraviolet radiation in the upper atmosphere, making it available to react with and destroy the ozone.

Chlorine was discovered in 1774 by the German chemist Karl Scheele, but English chemist Humphry Davy first proved it to be an element in 1810 and named it after its colour.

In ancient times, man knew of such chlorine-containing compounds as sodium chloride NaCl and ammonium chloride NH_4Cl. Later hydrochloric acid (HCl) became known and widely used. Numerous chlorine compounds were subjected to the scrutiny of researchers and there is no doubt that during manipulations with them, free chlorine was repeatedly obtained. Among those who observed free chlorine were such outstanding scientists as J. Glauber (of the Glauber's salt fame), J. Van Helmont, and R. Boyle. But even if this strange yellow-green gas had caught their attention, they would have hardly understood its nature.

The Swedish chemist C. Scheele was also mistaken. He prepared chlorine by the same method that is described in modern school textbooks: by the reaction of hydrochloric acid with manganese oxide (Scheele made use of ground pyrolusite that is natural MnO_2). It would be wrong to say that the scientist chose this method by chance. Scheele knew that the reaction of HCl with pyrolusite had to give rise as usual to inflammable air (known subsequently as hydrogen). Some gas was, indeed, liberated but it did not bear even remote likeness to inflammable air. It had a very unpleasant smell and an unpleasant yellow-green colour. The gas corroded corks and bleached flowers and plant leaves. The new gas proved to be a highly active chemical reagent. It reacted with many metals and, when with ammonia, formed a dense smoke (ammonium chloride NH_4Cl), its solubility in water was poor. Scheele did not utter the words "a new chemical element", although he had the discovery within his grasp and could follow the logical chain of arguments about its elementary nature. A zealous follower of the phlogistic theory, the Swedish chemist identified the gas discovered by him with hydrochloric acid that had lost phlogiston. He named it "dephlogisticated hydrochloric acid or dephlogisticated muric acid" (HCl was named muric acid after the Latin muria, "brine, salt water"). At that time, Scheele shared the opinion of H. Cavendish and other scientists that inflammable air (hydrogen) was actually phlogiston. It followed that the new gas had to be a simple substance (hydrochloric acid minus phlogiston) but Scheele did not make such seemingly obvious conclusion. Although 1774 is considered to be the new gas' date of discovery, much time was to pass before its nature was properly understood.

A. Lavoisier overturned the phlogistic theory. Even the name "dephlogisticated muric acid" evoked a strong protest in him. In his opinion, the acid obtained by Scheele was a compound of muric (hydrochloric) acid and oxygen. Oxidized muric acid - that is how Lavoisier named what we know as elemental chlorine now. The French chemist believed that all acids must contain oxygen combined with some element. Lavoisier called this element "murium" in the case of muric acid and included it into his "Table of Simple Bodies" (murium radical - radical muriatique).

The result was paradoxical; trying to elucidate the nature of the gas discovered by Scheele, Lavoisier only complicated the issue. Probably, this development in the history of chlorine was simply inevitable in the light of new theoretical conceptions. Some chemists attempted to prepare free murium but the attempts were fruitless and the nature of the new gas did not become clearer.

In 1807 H. Davy tried to solve the problem, subjecting the notorious muric acid to various manipulations. He attempted to decompose it electrolytically, but no decomposition was observed. No matter how ingeniously he treated oxymuric acid, he could not succeed in preparing water or liberating oxygen. In a word, the acid behaved as if it were a simple substance.

Moreover, its action on metals or their oxides yielded typical salts. Nothing else was left to Davy but to recognize that oxymuric acid consisted of only one simple substance, i.e. to recognize the elemental nature of the gas discovered more than 30 years earlier by Scheele. He reported on this to the Royal Society on November 19, 1810.

Davy proposed to name the element "Chlorine" from the Greek chloros meaning "yellow-green". Two years later, in 1812, the French chemist Gay Lussac proposed to change the name for "Chlor" (which became generally accepted except in English-speaking countries).

Gay Lussac in cooperation with Thenard began to study oxymuric acid almost simultaneously with Davy; at first, they wanted to prove that it was oxygen-free. The two scientists passed the acid through a red-hot porcelain tube over charcoal. If there had been oxygen in the gas discovered by Scheele, it would have been absorbed by charcoal. Although the composition of the gas at the inlet and outlet of the tube remained unchanged, this experiment did not shake the belief of the firm followers of A. Lavoisier about the composition of oxymuric acid.

Nevertheless, Davy's experiments strongly impressed the contemporary scientific community which gradually came to the conclusion that murium was in fact chlorine. In 1813 Gay Lussac and Thenard agreed with Davy. Only Berzelius for a long time continued to doubt the elemental nature of chlorine but in the end he also had to accept the truth. The elemental nature of chlorine became an irrefutable fact only after the discovery and study of iodine and bromine.

In 1811 the German chemist I. Schweiger proposed to name chlorine a "halogen" (from the Greek for "Salt" and "produce", i.e; "Salt-producing" because of its ability to combine readily with alkaline metals. At the time, the name was not accepted but later it became common for the group of similar elements:

fluorine, chlorine, bromine and iodine. Chlorine was obtained for the first time in liquid form in 1823 by M. Faraday.

IODINE

Iodine (Greek Iodes "violet") - greyish-black, non-metallic element, symbol I, atomic number 53, relative atomic mass 126.9044. It is a member of the halogen group. Its crystals give off, when heated, a violet vapour with an irritating odour resembling that of chlorine. It only occurs in combination with other elements. Its salts are known as iodides, which are found in sea water. As a mineral nutrient, it is vital to the proper functioning of the thyroid gland, where it occurs in trace amounts as part of the hormone thyroxine. Iodine is used in photography, in medicine as an antiseptic and in making dyes.

Its radioactive isotope ^{131}I (half-life of eight days) is a dangerous fission product from nuclear explosions and from the nuclear reactors in power plants, since, if ingested, it can be taken up by the thyroid and damage it. It was discovered 1811 by French chemist Bernard Courtois (1777-1838).

Iodine was the second halogen to be obtained in a free state. Both the appearance and chemical properties of iodine are rather peculiar. Was it the only halogen in existence, chemists would have to think hard about its nature, but the elemental chlorine had already been known and this fact helped to understand the nature of iodine.

B. Courtois, an entrepreneur from the French town of Dijon, was engaged, among other things, in the production of potash and saltpetre. He used ash of sea algae as the initial raw material. A mother solution of sea algae was formed under the action of water on the ash. Today, we know that the ash contains chlorides, bromides, iodides, carbonates and sulphates of some alkali and alkaline-earth metals. However, when

Courtois performed his experiments it was only known that the ash contained potassium and sodium compounds (chlorides, carbonates and sulphates). Upon evaporation, first, sodium chloride precipitated and then potassium chloride and sulphate. The residual mother solution contained a complex mixture of various salts, including sulphur-containing ones.

To decompose these sulphur compounds, Courtois added sulphuric acid to the solution. One day it so happened that he added a greater amount of acid than was necessary. Suddenly something unexpected happened: amazingly beautiful clouds of violet vapour appeared whose magnificence was marred only by their unpleasant, even lachrymose smell. Then followed something even more surprising: on the surface of cold objects the vapour did not condense forming heavy drops of a violet liquid but precipitated at once as dark crystals with metallic lustre. Courtois discovered many other interesting and unusual properties of the new substance. He had every reason to announce the discovery of a new chemical element but, evidently, the researcher was not confident enough and his laboratory was too poorly equipped to perform further investigations. He, therefore, turned for help to his friends, Ch. Desormes and N. Clement, asking them for permission to continue his experiments in their laboratory. He also asked them to report his discovery in a scientific journal.

Consequently, the report about "The Discovery of a New Substance Obtained from an Alkali Salt by Mr. Courtois" signed by N. Clement and Ch. Desormes appeared only in 1813 in the "Annales de chimie et de physique", i.e. two years after the discovery of the element. To enable other chemists to investigate the substance, B. Courtois gave a very small amount of it to a pharmaceutical firm in Dijon. Clement himself prepared a certain amount of iodine, studied its properties and was, probably, the first to advance an opinion that iodine resembled chlorine In 1813 J. Gay Lussac and H. Davy independently of each other proved the elemental nature of iodine. The French chemist suggested the name "iode" for the new element (from the Greek iodes meaning "violet colour") and the English scientist suggested the name "iodine". The first name found acceptance in the Russian language.

Iodine is a rare example of a chemical element whose properties were studied thoroughly during a short period of time after its discovery. Here a great contribution was made by Gay Lussac who even wrote a book on iodine which was in effect the first monograph in the history of science completely devoted to one element.

But the subsequent generations did not forget B. Courtois' contribution. A street in Dijon is named after him; this honour was bestowed on very few discoverers of chemical elements.

BROMINE

Bromine (Greek bromos 'stench') - dark, reddish brown, non-metallic element, a volatile liquid at room temperature, symbol Br, atomic number 35, relative atomic mass 79.904. It is a member of the halogen group, has an unpleasant odour and is very irritating to mucous membranes, its salts are known as bromides.

Bromine was formerly extracted from salt beds but is now mostly obtained from sea water, where it occurs in small quantities. Its compounds are used in photography and in the chemical and pharmaceutical industries.

This element, unusual in many respects, was the last of the natural halogens to be discovered (if, of course, we accept the discovery of fluorine by Scheele in 1771).

On an autumn day in 1825, the following event took place in the laboratory of L. Gmelin, a professor of medicine and chemistry at Heidelberg University. A student by the name of

C. Lowig brought to his teacher a thick-walled flask with an evil-smelling reddish brown liquid. Lowig told Gmelin that in his native town of Kreiznach he had studied the composition of water from a mineral spring. Gaseous chlorine turned the mother solution red. Lowig extracted with ether the substance that caused the colouring of the solution. It was a reddish brown liquid known subsequently as bromine.

Gmelin showed great interest in his student's work and advised him to prepare the new substance in greater amounts and to study its properties in detail. It was a reasonable piece of advice since Lowig had little experience as an experimenter; but the work required time and time factor turned against the student.

While he was assiduously preparing new portions of bad smelling reddish brown liquid, a large article appeared in the Annales de chimie et de physique. The article was entitled "Memoir on a Specific Substance Contained in Sea Water" and was written by A. Balard. He was a laboratory assistant at a pharmaceutical school in the French town of Montpellier. The properties of his "Specific substance" turned out to be quite similar to those of the reddish brown liquid obtained by Lowig. A. Balard wrote that in 1824 he began to study vegetation of salt marshes. He subjected marsh grasses to the action of various chemical reagents trying to extract useful compounds from them. He prepared a mother solution which turned brown under the action of some reagents, such as chlorine. Then A. Balard studied an alkaline solution obtained after the treatment of sea algae ash. As soon as chlorine water and starch were added to the solution, it separated into two layers. The lower part was blue and the upper one, reddish brown. A. Balard decided that the lower layer contained iodine which coloured starch blue. And what about the upper layer? Balard assumed that it contained a compound of chlorine with iodine. He tried to extract it but in vain. Only after that did the laboratory assistant from Montpellier dare to think that reddish brown colouring was caused by a new chemical element. Balard separated the reddish brown liquid, which was similar to that separated several months before by the unknown student Lowig who later became an academician and professor at Sorbonne.

Balard gave the new element a prosaic name "muride" from the Latin muria for "brine". He had an equally prosaic view of the nature of the element believing it to be the only non-metal liquid at room temperature like metallic mercury, which is liquid under the same conditions.

Balard's article did not remain unnoticed but, nevertheless, his friends advised him to send a report to the Paris Academy of Sciences. Balard followed the advice and on November 30, 1825, he sent a communication "Memoir on a Specific Substance Contained in Sea Water". The most important thing in the communication was the observation on similarity of muride with chlorine and iodine. The members of the Academy did not take such reports on trust and a special committee was set up to check Balard's experimental results. The committee, consisting of Gay Lussac, Vauqueline and Thenard, confirmed all the results obtained by Balard and only the name of the new element caused objections. The committee named it "bromine" from the Greek bromos which means "Stinking".

The committee made its ruling on August 14, 1826; the discovery of bromine was extremely important for chemistry.

And only one scientist met the news of the discovery with irritation. He was J. Liebig. Several years earlier he had received a bottle with a liquid from a German firm that asked Liebig to identify the liquid. The scientist did not analyse it thoroughly and made a hasty conclusion that the liquid was a compound of iodine with chlorine. When Liebig learnt about Balard's discovery he analysed the liquid remaining in the bottle and established that it was bromine. His contemporaries reported that Liebig said in temper: "It is not Balard who discovered bromine but bromine that discovered Balard".

Significance of Halogens for the Development of Chemistry

When determination of atomic masses (weights) became sufficiently accurate, the elements were arranged in accordance with increasing atomic weights. This made it possible to follow a change in chemical properties when passing from light to heavy elements and prepared the ground for discovering the periodic law. The concept of the natural groups was formed which combined chemically-similar elements. The triad chlorine-bromine-iodine was one of the first of such groups. It was thoroughly studied by the German chemist J. Dobereiner who can be regarded as one of Mendeleev's predecessors. An interesting fact was noted in this triad: the atomic weight of the middle element is half the sum of the atomic weights of the end elements. The same proved to be true for other triads (natural groups) of elements. The three chemical elements - chlorine, bromine and iodine - played their roles in the history of chemistry as the first "bricks" in the periodic law edifice.

Invaluable is the significance of these elements in the understanding of the composition and properties of acids. Initially, the discovery of chlorine supported the idea that all acids contained oxygen. At the later stages, chlorine was the first element for which both oxygen-containing and oxygen-free (hydrochloric) acids were obtained. Studying oxygen-containing acids of halogens, chemists gained new insights into the concepts of the strength of the acids and the degree of their dissociation. A comparison of the properties of hydrohalic acids was particularly fruitful and this does not exhaust the effect of the studies of chlorine, bromine and iodine on theoretical chemistry.

We see a similar picture in experimental chemistry. Halogenated hydrocarbons are the most important intermediates for preparing many organic compounds. This fact facilitated fast progress in organic syntheses in the 19th century. The chlorination method is widely used for extracting various valuable metals from minerals and ores; iodides are used for preparing extremely pure metals. Fluorine chemistry has become an independent branch of science.

BORON

Boron - non-metallic element, symbol B, atomic number 5, relative atomic mass 10.811. In nature it is found only in compounds, as with sodium and oxygen in borar. It exists in two allotropic forms - brown amorphous powder and very hard, brilliant crystals. Its compounds are used in the preparation of boric acid, water softeners, soaps, enamels, glass and pottery glazes. In alloys it is used to harden steel. Because it absorbs slow neutrons, it is used to make boron carbide control rods for nuclear reactors. It is a necessary trace element in the human diet. The element was named by English chemist Humphry Davy, who isolated it 1808, from borax.

Borax has been known for over 2,000 years and is used as a mild antiseptic and detergent.

Boron fibres have low density yet high strength and find extensive use as reinforcement material in composites for use in aircraft and spacecraft.

People widely used borax, one of the boron compounds, back in the Middle Ages. Probably borax had been known much earlier; it was reported that in the first millenium A.D. borax was used for soldering metals. However, the composition of natural borax remained unclear for a long time. Boric acid was obtained for the first time in 1702 by the Dutch physician W. Homberg who heated borax with sulphuric acid. It was used in medicine as "Homberg's sedative salt". In 1747 the French chemist Th. Baron tried to determine the composition of borax. He found that it contained Homberg's salt and soda; he was quite right: now we know that borax is a sodium salt of boric acid ($Na_2B_4O_7$).

The name of Swedish chemist T. Bergman deserves mention in the early history of boron. He believed that Homberg's salt was most likely not a salt but a compound resembling acid. As a matter of fact, it was he who introduced the name "boric acid". The term "boric radical" was mentioned in Lavoisier's "Table of Simple Bodies" and meant boron oxide. However, twenty years had to pass before the new chemical element, boron, was discovered.

It so happened that boron was discovered by several scientists: the French chemists L. Thenard and L.J. Gay Lussac and the English chemist H. Davy. They named the new element "boron" and "boracium" (from the word "borax"). The method of preparing the new element was the same in both cases: reduction of boric acid with metallic potassium. Independent discovery of a new chemical element by several researchers within ten days was a unique event in the history of chemical elements. Gay Lussac and Thenard announced their discovery on June 21, 1808 and Davy on June 30. Clearly, the priority of the French chemists in this case was ephemeral, especially because of the fact that it was Davy's previous discovery (preparation of elemental potassium) that gave the means for the separation of free boron.

CADMIUM

Cadmium - soft, silver-white, ductile and malleable, metallic element, symbol Cd, atomic number 48, relative atomic mass 112.40. Cadmium occurs in nature as a sulphide or carbonate in Zinc ores. It is a toxic metal that, because of industrial dumping has become an environmental pollutant. It is used in batteries, electroplating and as a constituent of alloys used for bearings with low coefficients of friction. It is also a constituent of an alloy with a very low melting point.

Cadmium is also used in the control rods of nuclear reactors, because of its high absorption of neutrons. It was named in 1817 by the German chemist Friedrich Stromeyer (1776-1835) after Greek mythological character cadmus.

In 1817 F. Stromeyer, a lecturer of the Chair of Chemistry at Gottingen University (Medical Department) and the chief inspector of chemist's shops in Hanover, found that calcination of zinc carbonate, sold in chemist's shops, produced a yellow compound although neither iron nor lead impurities were discovered in it.

This remarkable fact interested Stromeyer and he decided to visit a pharmaceutical firm in Salzgitter where he observed the same phenomenon. This prompted the scientist to study zinc oxide in more detail. To his surprise, Stromeyer discovered that the colour which zinc oxide acquired was due to a strange metal oxide never observed before. The chemist succeeded in separating this oxide from zinc oxide and reducing it to the metal.

His method consisted in the following: he dissolved contaminated zinc oxide in sulphuric acid and passed hydrogen sulphide through the solution; then he filtered off and washed the mixture of sulphides and dissolved it in concentrated hydrochloric acid. The acid was removed by evaporating the solution to dryness. Having dissolved the residue in water, F. Stromeyer added a large amount of ammonium carbonate. Since carbonate of the new metal did not dissolve in the presence of ammonium carbonate, Stromeyer filtered the precipitate off, washed it and transformed it into oxide which he reduced to metal with charcoal upon heating. As a result, bluish grey metal was obtained. However, since Stromeyer had only three grams of this metal, he could not thoroughly study its properties. Only in 1818 did he succeed in investigating the new metal.

F. Stromeyer named the metal "Cadmia", in accordance with the method of its preparation (as a result of calcination of $ZnCO_3$) "Cadmia" is the Greek for natural $ZnCO_3$. Independently of F. Stromeyer but somewhat later, cadmium was discovered by W. Maissner and K. Kersten in Germany (1818). Stromeyer's priority was contested by the German physician K. Roloff who, by the way, was the first to pay attention to the strange behaviour of commercially available zinc oxide upon heating. K. Kersten suggested to name the new metal "melinum" because of the yellow colour of its sulphide. It was also proposed to name the new metal "Klaprothium" (in honour of M. Klaproth) or "unonium" (after the asteroid) but none of the names found acceptance.

LITHIUM

Lithium (Greek Lithos "stone") - soft, ductile, silver-white, metallic element, symbol Li, atomic number 3, relative atomic mass 6.941. It is one of the alkali metals, has a very low density (far less than most woods) and floats on water (specific gravity 0.57); it is the lightest of all metals. Lithium is used to harden alloys and in batteries; its compounds are used in medicine to treat manic depression.

Lithium was named in 1818 by Swedish chemist Jons Berzelius, having been discovered the previous year by his student John A. Arfwedson (1792-1841). Berzelius named it after "stone" because it is found in most igneous rocks and many mineral springs.

The fate of the lightest metal is outwardly uneventful. It was the third alkali metal to be discovered in nature. Its abundance on Earth is much less than that of sodium and potassium, its minerals are rare and, therefore, it came relatively later to man's attention.

At the very beginning of the 18th century, the prominent Brazilian scientist and statesman J. Andradae Silva was travelling in Scandinavia. A passionate mineralogist, he wanted to enrich his collection with new specimens. He had luck and found two new minerals which he named petalite and spodumene. J. Andradae Silva found the minerals at the island of Uto belonging to Sweden. Soon spodumene was found in other places but the existence of petalite was doubted until, in 1817, it was found in Uto for the second time.

Therefore, spodumene was the first to become the subject of investigation. M. Klaproth studied it but discovered nothing except alumina and silica. In a word, spodumene was a typical alumino silicate. But the total mass of the isolated components was 9.5 percent less than the mass of the initial sample, and Klaproth could not explain the reason for this considerable loss. Meanwhile, his compatriot I. Nepomuk Von Fux discovered by chance that a pinch of spodumene turned the burner flame red. The scientist did not try to find the reason for this phenomenon and that was a mistake, since he could have discovered a new element in spodumene.

The second discovery of petalite attracted attention to the mineral. L. Vauquelin found alkali in it, in addition to alumina and silica, but erroneously identified it with potash. W. Hizinger obtained interesting and suggestive results but had no chance to explain them since the same data had already been published by the Swedish chemist I. Arfvedson to whom the credit for discovering lithium went. J. Berzelius in his letter to A. Berthollet, the famous French chemist, on February 9, 1818, described this event in the following way. A new alkali, he wrote, was discovered by I. Arfvedson, a skillful young chemist, who had been working in his laboratory for a year. Arfvedson found the alkali in the ore discovered earlier by Andrada at the Uto mine and named petalite. The ore consisted of 80 percent silicon oxide, 17 percent aluminium, and 3 percent the new alkali. The conventional method used to extract the alkali consisted in heating the ground ore with barium carbonate and separating all earths from it.

Analysing petalite, Arfvedson from the very beginning discovered that the losses of the material amounted to about 4 percent. The Swedish chemist (like M. Klaproth in his time) tried to find the answer again and again, sweeping aside various assumptions and at last reached the truth - it was a new alkali of unknown nature. It was clear that this alkali was formed by a new alkali metal. I. Arfvedson asked his teacher to help him choose the name for the metal and the scientists decided to name it "Lithium" (from the Greek lithios for "Stone"). This name is a reminder that lithium was discovered in the mineral kingdom whereas two other alkali metals (sodium and potassium) in the plant kingdom.

Arfvedson published the report on the discovery of lithium in petalite in 1819 but already in April, 1818, the scientist found the new alkali metal in other minerals as well. The secret of spodumene, which Klaproth had failed to solve, was finally cleared - the mineral contained about 8 percent of lithium. And one more mineral, lepidolite, known for a long time, was also found to contain up to 4 percent of the lightest alkali metal.

The German chemist K. Gmelin observed lithium salts to turn the burner flame a beautiful shade of red (to I. Von Fux's great irritation).

By the late 1818 H. Davy succeeded in separating pure lithium, though in very small amounts. It became possible to obtain large amounts of lithium only in the late 1850s when the German chemists Bunsen and Matissen developed an industrial process of electrolysis of lithium chloride.

SELENIUM

Selenium (Greek selene 'Moon') - grey, non-metallic element, symbol Se, atomic number 34, relative atomic mass 78.96. It belongs to the sulphur group and occurs in several allotropic forms that differ in their physical and chemical properties. It is an essential trace element in human nutrition obtained from many sulphide ores and selenides, it is used as a red colouring for glass and enamel.

Because its electrical conductivity varies with the intensity of light, selenium is used extensively in photoelectric devices. It was discovered in 1817 by Swedish chemist Jons Berzelius and named after the Moon because its properties follow those of tellurium, whose name derives from Latin T.

Selenium is still another element that chemists had met long before its discovery, but failed to identify owing to its having been masked by the presence of other similar elements. Thus, selenium remained undiscovered, "hiding" behind sulphur and tellurium. Only in 1817, did it surrender to the Swedish chemists - the famous J. Berzelius and his assistant G. Gahn. Inspecting a sulphuric acid factory in Gripsholm on September 23, they found a small amount of a precipitate, partially red and partially light brown, in sulphuric acid. On heating in the flame of a blowpipe the precipitate emitted a weak smell of radish and transformed into a regulus with a leaden lustre. In Klaproth's opinion the smell of radish pointed to the presence of tellurium. Similar smell was noticed in the Falun mine where pyrite required for the acid production was extracted. Curiosity and hope to find this rare metal in the brown precipitate forced Berzelius to investigate it. However, he did not discover tellurium. Then he collected the deposits formed after several months of sulphur combustion for sulphuric acid production in the Falun factory and obtained a large amount of precipitate. Thoroughly analysing the precipitate, Berzelius came to the conclusion that it contained an unknown metal whose properties were similar to those of tellurium. By analogy, the new metal was named "selenium" from the Greek selenus for the Moon (as tellurium is named after our planet). Berzelius studied many properties of selenium and described them in an article "The Study of a New

Mineral Body Found in Sulphur Extracted in Falun" published in 1818 in the journal Annales de chimie et de physique.

SILICON

Silicon (Latin Silicium "silica") - brittle, non-metallic elements, symbol Si, atomic number 14, relative atomic mass 28.086. It is the second most abundant element (after oxygen) in the Earth's crust and occurs in amorphous and crystalline forms. In nature it is found only in combination with other elements, chiefly with oxygen in silica (silicon dioxide, SiO_2) and the silicates. These form the mineral quartz, which makes up most sands, gravels, and beaches.

Pottery glazes and glassmaking are based on the use of silica sands and date from prehistory. Today the crystalline form of silicon is used as a deoxidizing and hardening agent in steel and has become the basis of the electronics industry because of its semiconductor properties, being used to make "silicon chips" for microprocessors.

The element was isolated by Swedish chemist Jons Berzelius in 1823, having been named in 1817 by Scottish chemist Thomas Thomson by analogy with boron and carbon because of its chemical resemblance to these elements.

Silicon is the second most abundant element on Earth after oxygen. Although it constitutes 28 percent of the earth's crust, its abundance did not make for its early discovery. The reason for this lies in the difficulty of reducing silicon from its oxide.

Generally speaking, there is every ground to classify silicon as an element of antiquity. Its compounds were known and used from time immemorial (suffice it to mention silicon tools of primitive man). We classified carbon as an element of antiquity since it was known in a free state from very remote times. However, that carbon is a chemical element became clear only two hundred years ago. Glass, in the long run, is also a silicon material. However, the date of silicon discovery is the date of its preparation in a free state since such is the established practice in the history of science.

At the turn of the 18th century, many scientists believed that silica, or silica earth, contained an unknown chemical element and tried to isolate it in a free state. H. Davy attempted to decompose silica with an electric current - the method by which a number of alkali metals had already been prepared - but without success. The scientist's attempt to prepare free silicon by passing metallic potassium vapour over red-hot silicon oxide also failed. In 1811 L.J. Gay Lussac and L. Thenard applied themselves to the problem. They observed a vigorous reaction between silicon tetrafluoride and metallic potassium; a reddish brown compound was formed in the reaction. The scientist could not reveal the nature of the product; most likely, it was contaminated amorphous silicon.

At last, in 1823, J. Berzelius had a stroke of good luck. The Swedish chemist heated a ground mixture of silicon oxide, iron and charcoal to a high temperature and obtained an alloy of silicon and iron (ferrosilicium), the composition of which he was able to prove. To separate free silicon, J. Berzelius repeated L. Thenard and L.J. Gay Lussac's experiments and also obtained a brown mass. Under the action of water, hydrogen bubbles were liberated and free amorphous silicon was formed as a dark brown insoluble powder which contained potassium silicofluoride as an impurity. Berzelius removed the impurity by washing the precipitate for a very long time.

Another method proposed by J. Berzelius - calcination of potassium fluorosilicate with an excess of potassium - proved to be more successful and straightforward. The sintered mass was decomposed with water and as a result, pure amorphous silicon was obtained. J. Berzelius showed that upon calcination, silicon was transformed into silicon; this makes Berzelius the

117

discoverer of silicon. Crystalline silicon was obtained in 1854 by A. Saint Claire Deville during separation of metallic aluminium. The Latin name "Silicium" originates from "silex" meaning "a hard stone".

ALUMINIUM

Aluminium - lightweight, silver-white, ductile and malleable, metallic element, symbol Al, atomic number 13, relative atomic mass 26.9815. It is the third most abundant element (and the most abundant metal) in the Earth's crust, of which it makes up about 8.1% by mass. It is an excellent conductor of electricity and oxidizes easily, the layer of oxide on its surface making it highly resistant to tarnish. In USA, the original name suggested by the British Scientist Humphry Davy, "aluminum", is retained.

Because of its rapid oxidation, a great deal of energy is needed in order to separate aluminium from its ores and the pure metal was not readily obtainable until the middle of the 19th century. Commercially, it is prepared by the electrolysis of bauxite. In its pure state aluminium is a weak metal, but when combined with elements such as copper, silicon, or magnesium it forms alloys of great strength.

Because of its light weight (specific gravity 2.70), aluminium is widely used in the shipbuilding and aircraft industries. It is also used in making cooking utensils, cans for beer and soft drinks and foil. It is much used in steel-cored overhead cables and for canning uranium slugs for nuclear reactors. Aluminium is an essential constituent in some magnetic materials and as a good conductor of electricity, is used as foil in electrical capacitors. A plastic form of aluminium, developed in 1976, which moulds to any shape and extends to several times its original length, has uses in electronics, cars, building constructions and so on.

Aluminium sulphate is the most widely used chemical in water treatment worldwide, but accidental excess (as at Camelford, N Cornwall, England, July 1989) makes drinking water highly toxic and discharge into rivers kills all fish.

Aluminium is a chemical element to which history was unjust. The third most abundant metal on Earth after oxygen and silicon and found practically everywhere in the earth's crust (in 250 minerals, at least) aluminium was discovered only in 1825. And still, this later discovery of aluminium is not accidental. It was due to the extreme stability of aluminium oxide. To separate metallic aluminium from it is a tall order even in our times, to say nothing of the last century. Such reducing agents as charcoal and hydrogen could not separate the metal from the oxide. Only alkali metals, first of all potassium, made it possible "to capture the fortress". This shows how the discovery of some elements created the prerequisites for the discovery of others: free aluminium was first prepared with the help of potassium.

Man knew of various aluminium compounds in very remote times, Clay and brick are nothing but usual aluminosilicates. Alumina (aluminium oxide) was a constant companion of man but many centuries were required to prove the presence of a new metal in it. Aluminium is one of the main components in such precious stones known from time immemorial as ruby and garnet, sapphire and turquoise. Alums were known for a very long time. In Latin they were named alumen - the word which contained the root of the future "aluminium". However, the composition of alums remained undetermined for a long time and they were often confused with other compounds.

In 1754 the German chemist Marggraf tried to shed light on the problem. Having added pure alkali to the alum solution, he obtained a dense white precipitate which he named "alum earth". Then Marggraf observed that the addition of sulphuric acid to the "earth" yielded alum; thus, the composition of alum was established. And, finally, Marggraf demonstrated the presence of the "alum earth" in clays. Had history so willed it, Marggraf would have been acclaimed as the discoverer of this element, but history waited for somebody else to prepare pure aluminium. Only 30 years after Marggraf's experiment did it become clear that alumina was an oxide of an unknown element. This was suggested by A. Lavoisier who placed "alum earth" into his "Table of Simple Bodies". But no attempts were made for some time to separate the element in a free state.

The first attempt was made by H. Davy and J. Berzelius, who tried to decompose alumina with the aid of electric current, but in vain; it was only H. Davy's proposal (1807) to name the element "aluminium" that had any practical importance. This name became internationally accepted although in Russia the name "glinium" (from the Russian word for "Clay") was used for a long time.

The first who managed to obtain metallic aluminium was the Danish scientist H. Oersted known in history as a physicist rather than as a chemist. He discovered the induction of magnetic field of an electric current, but preparation of pure aluminium showed him to be also a skilful chemist. Having red-heated a mixture of alumina with charcoal, Oersted passed chlorine through it; as a result anhydrous aluminium chloride was obtained. Then the scientist heated the new compound with potassium amalgam and obtained amalgam of aluminium for the first time. As soon as Oersted distilled off the mercury, he discovered pieces of metal that looked like tin. The product contained impurities but, nevertheless, this was the birth of metallic aluminium. Oersted published an article in a little known Danish Journal which passed practically unnoticed in the scientific circles. And news of Oersted's achievement did not reach many chemists. Therefore, some historians believe that aluminium was discovered not by Oersted but F. Wohler.

The second discovery of aluminium took place two years later, in 1827. Undoubtedly, F. Wohler was a more skilful experimenter than Oersted and his process of separating pure aluminium was more sophisticated. At first Wohler's attempt to obtain the metal using the Danish Scientist's method failed but soon he succeeded in preparing small amounts of anhydrous aluminium chloride. Wohler developed his own procedure for the process:

1. Preparation of aluminium hydroxide;
2. Preparation of a thick paste from aluminium hydroxide, charcoal and vegetable oil;
3. Calcination of the paste and preparation of a mixture of aluminium with charcoal powder;
4. Preparation of pure anhydrous $AlCl_3$ by passing dry chlorine through the mixture. The complexity of this procedure was rewarded by the purity of the product. The scientist decomposed $AlCl_3$ with potassium under conditions ensuring the highest possible purity of the metal. F. Wohler was the first chemist to describe the most important properties of metallic aluminium and in 1845 he prepared aluminium in the form of an ingot.

However, Wohler, like his predecessors, did not obtain pure aluminium. The decisive word was said by the French chemist A. Saint Claire Deville. In 1854 he prepared the samples of pure metal, using sodium instead of potassium for the reduction stage. Simultaneously with Bunsen he performed electrolysis of melted double chloride of aluminium and sodium: this was the first instance of producing aluminium electrochemically. A. Saint Claire Deville also pioneered the development of an industrial process of aluminium production.

It is difficult to believe that only one hundred years ago this silvery metal was extremely expensive and was even called "Clay Silver". Things made of aluminium cost no less than gold ones. Only after the processes for producing cheap electric energy had been developed and rich deposits of aluminium ore had been found, did aluminium become a metal for everyday uses.

THORIUM

Thorium - dark-grey, radioactive, metallic element of the actinide series, symbol Th, atomic number 90, relative atomic mass 232.038. It occurs throughout the world in small quantities in minerals such as thorite and is widely distributed in monazite beach sands. It is one of three fissile elements (the others are uranium and plutonium) and its longest-lived isotope has a half-life of 1.39×10^{10} years. Thorium is used to strengthen alloys. It was discovered by Jons Berzelius in 1828 and was named by him after the Norse god Thor.

In 1815 J. Berzelius, the discoverer of the element, named it thorium in the honour of Thor, the ancient Scandinavian god of thunder. But the famous Swedish chemist anticipated the events: no new element was discovered by him that year. He analysed a rare mineral from Falun mines in which he discovered what he believed to be the oxide of an unknown element. Berzelius thought this justified the addition of one more name to the list of the existing elements. No contemporary dared even to doubt the discovery since in those times the scientists had boundless trust in Berzelius. However, Berzelius himself had doubt and justifiably so: ten years later it was shown that the oxide observed by him was yttrium phosphate (yttrium had already been known for a long time). Thus, in 1825 the past triumph turned sour.

A year later, F. Wohler reported the discovery of a new element in a rare Norwegian mineral now known under the name of "Pyrochlore". Wohler did not attach particular importance to this observation and as it turned out, mistakenly so.

Meanwhile, G. Esmark found a heavy black mineral on the Leven Island near the shores of Norway. The scientist sent a sample of the mineral to J. Berzelius who thoroughly analysed it. In 1828 Berzelius reported isolation of silicates of a new element from mineral. The old name "thorium" proved useful. The mineral which had become the source of thorium-2 was named by J. Berzelius "thorite".

When Berzelius studied the properties of thorium, Wohler paid attention to the fact that they were similar to those of the element which he had left without attention in 1826. Wohler was much more disappointed when six years later the famous German scientist and traveller W. Humboldt presented him with a sample of pyrochlore from Siberia. Wohler discovered thorium in it as a few years earlier, he had found it in the Norwegian pyrochlore. Thus, thorium played a trick on Wohler.

J. Berzelius tried to separate pure thorium but in vain. For very long the element was known in the form of its oxide and only in the 1870's was it prepared in the metallic form. Thus, thorium became the second radioactive element (after uranium) to be discovered by the conventional chemical analysis having nothing to do with radioactivity.

VANADIUM

Vanadium - silver-white, malleable and ductile, metallic element, symbol V, atomic number 23, relative atomic mass 50.942. It occurs in certain iron, lead and uranium ores and is widely distributed in small quantities in igneous and sedimentary rocks. It is used to make steel alloys, to which it adds tensile strength.

Spanish mineralogist Andre's del Rio (1764-1849) and Swedish chemist Nils Sefstrom (1787-1845) discovered Vanadium independently, the former in 1801 and the latter in 1831. Del Rio named it "erythronium", but was persuaded by other chemists that he had not in fact discovered a new element; Sefstrom gave it its present name, after the Norse goddess of love and beauty, Vanadis (or Freya).

Long, long ago there lived in the Far North Vanadis, a beautiful goddess. One day when she was reclining comfortably in her chair she heard a knock on the door. She thought to herself: "Let him knock once more." But the knock was not repeated and she heard someone go away. The Goddess was curious: "Who could that modest and diffident visitor be?" She opened a window and looked out. That was old Wohler himself who, of course, would have deserved a reward if he had been more persistent.

A few days later she again heard knocking on the door but this time it went on and on until she opened the door. She was confronted by Nils Sefstrom. They fell in love with each other and had a son whom they named Vanadium. That was the name of the new metal.

This is how the Swedish chemist Berzelius described the history of its discovery in his letter to F. Wohler on January 28, 1831. The story was rather unusual and not the least role in it was played by the ability of Vanadium to form salts of varied colours.

In Mexico, near the village of Cimapan, deposits of lead ore were found and in 1801 a sample fell into the hands of Andres Manuel del Rio, a professor of mineralogy from Mexico City. The scientist, a good analyst, studied the sample and came to the conclusion that it contained a new metal similar to chromium and uranium. Del Rio obtained several compounds of the metal which were all of different colours. The scientist named it "panchromium", the Greek for omnicoloured, but subsequently changed it into "erytronium" which means "red" since many salts of the new element turned red upon heating. The name of Del Rio was little known to European chemists who, learning about his results, doubted them. The Mexican mineralogist himself lost confidence and studying "erytronium", he practically "closed" his discovery saying that the element was nothing else than lead chromate. He sent a new article to Europe entitled "The Discovery of Chromium in Lead Ore from Cimapan". H. Collet-Descoties from Paris analysed a sample of the ore in 1809 and confirmed the Mexican scientist's erroneous conclusion. Erroneous, because Del Rio had really discovered Vanadium. It is difficult to explain why Del Rio was so unsure of his results. In 1832 after the second discovery of Vanadium he wrote in a text-book on mineralogy that the metal discovered by him was Vanadium and not chromium. But the credit for discovering Vanadium went to the Swedish chemist N. Sefstrom.

It was Sefstrom who in 1830 isolated a small amount of the new element from the iron ore extracted in the Taberg mine. Shortly before the discovery of the new element, F. Wohler studied the lead ore from Cimapan in which thirty years before A. del Rio had found "erytronium". Wohler wrote to J. Liebig on January 2, 1831, that he had already found something new in the ore. However, experimenting with hydrogen fluoride vapour, Wohler was poisoned and could not work for several months. One can imagine how disappointed he was when he learned about N. Sefstrom's discovery. J. Berzelius tried to console his friend and colleague, writing to him that a chemist who had discovered a method of synthesizing an organic compound (Wohler synthesized urea) could well renounce a claim to the priority of discovering a new element since his accomplishment was equivalent to the discovery of ten new elements. J. Berzelius and N. Sefstrom named the new element "Vanadium" after Vanadis, the Scandinavian goddess of beauty.

Meanwhile Wohler ended the study of the Mexican ore and proved that it contained Vanadium and not chromium as A. del Rio believed. Subsequently, this mineral was named "Vanadinite"; it was found in different parts of the globe. J. Berzelius and N. Sefstrom continued to study Vanadium and concluded that it was similar to chromium. Their attempts to prepare metallic Vanadium were unsuccessful and for some time, it seemed that they mistook either oxide or nitride of Vanadium for the metal. The final chapter in the Vanadium story brings up the name of the English chemist H. Roskoe. In 1860's he performed a detailed study of the chemical properties of Vanadium and showed that this element was similar neither to chromium nor uranium. On the contrary, he thought that Vanadium was similar to niobium and tantalum on the one hand and to the elements of the phosphorus group on the other. In 1869 Roskoe succeeded in preparing metallic Vanadium. D.I. Mendeleev highly appreciated the work of this scientist believing that it had played a great role in the discovery of the periodic law.

5

Elements discovered by the electrochemical method

This short chapter deals with the discovery of two alkali metals, sodium and potassium and two alkaline-earth metals, magnesium and calcium. They were discovered, directly in a free state, in the first decade of the 19th century. The compounds of these metals had been known from very remote times and it is hardly possible to establish more or less accurately when common salt, potash, lime or magnesia came into use. All these compounds had been man's companions long before the metals contained in them were discovered.

A. Lavoisier included lime and magnesia into "The Table of Simple Bodies" but excluded potassium and sodium hydroxides believing that they had complex composition and their nature had to be further studied. One might say that history was unjust to these elements, for barium, for instance, was isolated in a metallic state simultaneously with them, but had been discovered much earlier. However, history is a wayward lady. The discovery of sodium, potassium, magnesium and calcium is interesting in that it was made possible by electric current being successfully used for the first time. This marked the birth of the electrochemical method, a subsidiary to the chemical analysis. Subsequently, electrolysis of melted compounds made it possible to obtain other metals discovered earlier in their compounds.

That is why we considered it justified to devote a separate chapter to the history of sodium, potassium, magnesium and calcium. The time span in question is two years and H. Davy, one of the founders of electrochemistry, is the main character.

SODIUM AND POTASSIUM

Sodium - soft, waxlike, silver-white, metallic element, symbol Na (from Latin natrium), atomic number 11, relative atomic mass 22.898. It is one of the alkali metals and has a very low density, being light enough to float on water. It is the sixth most abundant element (the fourth most abundant metal) in the Earth's crust. Sodium is highly reactive, oxidizing rapidly when exposed to air and reacting violently with water. Its most familiar compound is sodium chloride (common salt), which occurs naturally in the oceans and in salt deposits left by dried-up ancient seas.

Other sodium compounds used industrially include sodium hydroxide (caustic soda, NaOH), Sodium Carbonate (washing soda, Na_2CO_3) and hydrogencarbonate (Sodium bicarbonate, $NaHCO_3$) Sodium nitrate (Saltpetre, $NaNO_3$, used as a fertilizer) and sodium thiosulphate (hypo, $Na_2S_2O_3$ used as a photographic fixer). Thousands of tons of these are manufactured annually. Sodium metal is used to a limited extent in spectroscopy, in discharge lamps and alloyed with potassium as a heat-transfer medium in nuclear reactors. It was isolated from caustic soda in 1807 by English chemist Humphry Davy.

Potassium (Dutch Potassa "potash") - soft, waxlike, silver-white, metallic element, symbol K (Latin Kalium), atomic number 19, relative atomic mass 39.0983. It is one of the alkali metals and has a very low density - it floats on water and is the second lightest metal (after lithium). It oxidizes rapidly when exposed to air and reacts violently with water of great abundance in the Earth's crust, it is widely distributed with other elements and found in salt and mineral deposits in the form of potassium aluminium silicates.

Potassium, with sodium, plays a role in the transmission of impulses by nerve cells and so is essential for animals; it is also required by plants for growth. The element was discovered and named in 1807 by English chemist Humphry Davy, who isolated it from potash in the first instance of a metal being isolated by electric current.

Man had known sodium and potassium compounds for a very long time. Carbonates of these metals were used in Egypt for laundry. Common salt, one of the most widespread sodium compounds, was used in foods from time immemorial; in some countries it was very expensive and sometimes wars were waged for the right to possess salt mines. Sodium carbonate was usually obtained from salt lakes whereas potassium carbonate by leaching plant ash; for this reason the former was named mineral alkali and the latter vegetable alkali. The word "alkali" was introduced by Geber, a medieval alchemist, although he made no distinction between the two carbonates. The differences in their nature were first mentioned in 1683. The Dutch scientist I. Bon noted that when soda and potash were used in the similar process, the shapes of the precipitated crystals were different depending on the initial product.

In 1702 G. Stahl noted the difference in crystals of some sodium and potassium compounds. This was an important step in distinguishing between soda and potash. In 1736 the French chemist A. de Monsean proved that soda was always present in common salt, Glauber's salt and in borax. Since an acidic constituent of soda was known, the nature of the basic constituent was of great interest. According to Monsean, soda formed Glauber's salt with sulphuric acid, cubic saltpetre (Sodium nitrate) with nitric acid and a variety of sea salt with hydrochloric acid: isn't this reason enough to deduce that soda is the basis of sea salt?

Although chemists had suspected for a long time that alkali earths were oxides of metals, the nature of soda and potash had not been studied up to the early 19th century. Even Lavoisier had no definite idea on this subject. He did not know what the basic constituents of soda and potash were and assumed that nitrogen could be a constituent. This confusion seems to stem from the similarity between the properties of sodium, potassium and ammonium salts.

Credit for determining these constituents belongs to H. Davy. At first he was dogged by failures: he could not separate metals from soda and potash with the aid of a galvanic battery. However, soon the scientist understood his error - he used saturated aqueous solutions but the presence of water hinders decomposition. In October, 1807, Davy decided to melt anhydrous potash and as soon as he started electrolysis of the alkali hydroxide melt, small balls resembling mercury with bright metallic lustre appeared on the negative electrode immersed into the melt. Some of the balls burnt up immediately with an explosion forming bright flame while the others did not burn, but just dimmed and became covered with a white film. Davy concluded that numerous experiments had shown that the balls were the substance which he had been looking for and this substance was highly inflammable potassium hydroxide.

Davy studied this metal thoroughly and found that when it reacted with water, the resulting flame was due to burning of the hydrogen liberated from water. Having studied the metal obtained from potassium hydroxide, H. Davy began to search for sodium hydroxide using the same method and he succeeded in separating another alkali metal. The scientist noted that for its preparation a much more powerful battery was required than in the experiments with potash. Nevertheless, the properties of both metals turned out to be similar.

For a short time the scientist carefully studied the properties of potassium and sodium. Some chemists doubted the elemental nature of sodium and potassium believing that they were compounds of alkalis with hydrogen. However, Gay Lussac and Thenard proved convincingly that Davy had, indeed obtained simple substances.

MAGNESIUM

Magnesium - lightweight, very ductile and malleable, silver-white, metallic element, symbol Mg, atomic number 12, relative atomic mass 24.305. It is one of the alkaline-earth metals and the lightest of the commonly used metals. Magnesium silicate, carbonate and chloride are widely distributed in nature. The metal is used in alloys and flash photography. It is a necessary trace element in the human diet and green plants cannot grow without it since it is an essential constituent of the photosynthetic pigment chlorophyll ($C_{55}H_{72}MgN_4O_5$).

It was named after the ancient Greek city of Magnesia, near where it was first found. It was first recognized as an element by Scottish chemist Joseph Black in 1755 and discovered in its oxide by English Chemist Humphry Davy in 1808. Pure magnesium was isolated in 1828 by French chemist Antoine-Alexandre-Brutus Bussy.

Magnesium compounds such as asbestos, talcum, dolomite and nephrite have been known from very remote times and used for various purposes. They, however, were not recognized as individual substances but were considered to be varieties of lime.

In 1618 H. Wiker found mineral springs near Epsom in England. In 1695 a salt (magnesium sulphate) with a bitter taste was discovered in the Epsom spring water and later it was used in medicine.

Scientists established that artificial Epsom salt could be prepared by adding sulphuric acid to the mother solution remaining after the purification of salt extracted from sea water. The difference between Epsom and Glauber's (sodium sulphate) salts was established but the difference between lime and white magnesia remained unclear for a long time. J. Black was the first to establish the different solubilities of these compounds and their sulphates in water. According to C. Newman, magnesium oxide was considered to be white magnesia in contrast to black magnesia, which is pyrolusite.

Metallic magnesium (although not very pure and in a very small amount) was obtained for the first time in 1808 by H. Davy who used the same procedure as that for isolating potassium and sodium. Large amounts of the pure metal were obtained in 1831 by the French chemist A. Bussy. The name of the element is derived from the word "magnesia".

CALCIUM

Calcium (Latin calcis "lime") - soft, silvery-white metallic element, symbol Ca, atomic number 20, relative atomic mass 40.08. It is one of the alkaline-earth metals. It is the fifth most abundant element (the third most abundant metal) in the Earth's crust. It is found mainly as its carbonate $CaCO_3$, which occurs in a fairly pure condition as chalk and limestone (calcite). Calcium is an essential component of bones, teeth, shells, milk and leaves and it forms 1.5% of the human body by mass.

Calcium ions in animal cells are involved in regulating muscle contraction, hormone secretion, digestion and glycogen metabolism in the liver.

The element was discovered and named by the English chemist Humphry Davy in 1808. Its compounds include slaked lime (calcium hydroxide, $Ca(OH)_2$); plaster of paris (Calcium sulphate, $CaSo_4.H_2O$); Calcium Phosphate $(Ca_3(PO_4)_2)$, the main constituent of animal bones; calcium hypochlorite $(CaOCl_2)$, a bleaching agent; calcium nitrate $(Ca(NO_3)_2.H_2O)$ a nitrogenous fertilizer; calcium carbide (CaC_2), which reacts with water to give ethyne (acetylene); Calcium cyanamide $(CaCN_2)$, the basis of many pharmaceuticals, fertilizers and plastics, including melamine; calcium cyanide $(Ca(CN)_2)$, used in the extraction of gold and silver and in electroplating and others used in baking powders and fillers for paints.

Many calcium minerals, for instance, limestone, gypsum, alabaster, that is, mainly, carbonate and sulphate minerals, have been known for a very long time. In the old days people already knew how to transform limestone into lime by calcination, as was reported by Pliny the Elder. However, it was only in 1755 that J. Black showed that the weight (mass) losses during calcination were completely caused only by the removal of fixed air. i.e. carbon dioxide.

The name "alabaster" served in antiquity to denote two minerals. For one of them (a variety of calcium sulphate) the name survived up to our days, but in Egypt, for example, "alabaster" meant a variety of calcite (calcium carbonate).

Gypsum has also been used from times immemorial as a construction material. Gypsum-based solutions found application in building pyramids, temples and other edifices. Theophrastos applied the name "gypsum" to two minerals: gypsum itself and the product of its partial dehydration. Pure calcium oxide was described by the German chemist I. Pott back in 1746; however, attempts to obtain metal from it with the aid of various reducing agents failed.

The right approach was suggested by H. Davy. First, he attempted to obtain calcium by passing electric current through humid earth insulated from the air by a kerosene layer. (In a similar way he had tried to prepare barium and strontium).

As a result of his experiments, Davy developed the following method of preparing pure alkaline-earth metals. He mixed humid earth with 1/3 (by mass) of mercury oxide and placed the mixture into a platinum vessel connected to the positive pole of a high-voltage battery. Then he introduced a drop of mercury at the centre of the mixture. The platinum electrode placed in the drop was connected with the negative pole of the battery. Amalgam obtained in this way was then separated into mercury and silvery white metal, calcium. Davy prepared pure calcium in 1808. In the same year J. Berzelius and M. Pontin obtained calcium independently of Davy using a similar method. The name of the element originates from the Latin word calx, which means "lime".

6

Elements discovered by the spectroscopic method

Hardly a decade passed in the 19th century without additions to the list of chemical elements, sometimes considerable additions. The year 1850 is the only exception; not a single new element was discovered during this period. This is hardly strange: analytical chemistry had already done everything in its power. By the middle of the century, chemical analysis made it possible to discover all the elements whose discovery did not demand other fine techniques. The discovered elements were either sufficiently abundant in a native state or scientists were lucky to find minerals containing rare elements. By the mid-19th century about 60 elements were already known.

This full in the history of discoveries of new elements was ended by the spectral method developed in 1859-1860 by the German scientists R. Bunsen and G. Kirchhoff. And at once reports appeared about the discovery of new elements, which announced themselves via new spectral lines. Four chemical elements (cesium, rubidium, thallium and indium) came to light owing to the spectroscopic method.

CAESIUM

Caesium (Latin caesius "bluish-grey") - soft, silvery-white, ductile metallic element, symbol Cs, atomic number 55, relative atomic mass 132.905. It is one of the alkali metals and is the most electropositive of all the elements. In air it ignites spontaneously and it reacts vigorously with water. It is used in the manufacture of photoelectric cells. The name comes from the blueness of its spectral line.

The rate of vibration of caesium atoms is used as the standard of measuring time. Its radioactive isotope Cs-137 (half-ife 30.17 years) is a product of fission in nuclear explosions and in nuclear reactors; it is one of the most dangerous waste products of the nuclear industry, being a highly radioactive biological analogue for potassium.

Caesium, a rare alkaline-earth metal, was fated to become the first chemical element whose presence on Earth was established by spectroscopy, although its fate could have been different. Back in 1846 the mineralogist A. Breithaupt, studying minerals and ores from the island of Elba, noted a coloured variety of quartzite, which he named pollux (or pollycite). The sample of pollux then fell into the hands of the German chemist K. Plattner from Freiberg, a professor of metallurgy in the Mining Academy. Plattner had a minute amount of pollux sufficient only for one analytical experiment. Having separated the constituents of the mineral and finding nothing new, Plattner, to his surprise, noted that the sum total of the constituents was only 92.75 percent. The reason for this remained unclear since Plattner had no pollux left. The scientist, however, established the following: Pollux had the highest alkali content among all known silicates. It is now clear that caesium was safely masked by the much larger amounts of sodium and potassium and Plattner was not able to extract it.

In 1860 R. Bunsen and G. Kirchhoff studied the chemical composition of various mineral spring waters by spectroscopy. After the separation of calcium, strontium, magnesium and lithium from a sample of Durkheim mineral water, a drop of the evaporated solution was studied spectroscopically. The scientists observed two pronounced blue lines close to each other. One of them almost coincided with the strontium line. Bunsen and Kirchhoff asserted that since no substance was known to have such spectral lines it had to be an unknown substance, belonging to the group of alkali metals. They proposed to name it "Caesium" (symbol Cs) from the Latin Caesius: in ancient times this word was used to describe the blueness of the upper part of the firmament. The beautiful blue vapour of caesium helped to prove the presence of a few millionths of a milligram of this substance in a mixture with sodium, lithium and strontium.

On April 11, 1860, R. Bunsen wrote to G. Roskoe (his collaborator in a study in photochemistry) about his investigation of the new alkali metal. On May 10 he reported the discovery of caesium to the Berlin Academy of Sciences. Six months later Bunsen already had 50g of almost pure caesium chloroplatinate. To obtain such an amount of the product, it was required to process nearly 300 tons of mineral water; about one kilogram of lithium chloride was isolated as a side product. These figures show how small was the caesium content in mineral spring waters.

Four years later, the Italian analyst F. Pizani set to study pollux, earlier investigated by Plattner. Pizani had a stroke of luck; he discovered caesium in the mineral and demonstrated that the German scientist had erroneously taken caesium sulphate for a mixture of sodium and potassium sulphates. Pure caesium, however, was separated only in 1882 by the German chemist K. Satterberg via electrolysis of a mixture of cyanides CsCN and $Ba(CN)_2$. In Russia, Beketov prepared caesium almost at the same time and independently of Satterberg by reducing caesium aluminate $(CsAlO_2)$ with magnesium in a hydrogen flow.

RUBIDIUM

Rubidium (Latin rubidus "red") - soft, silver-white, metallic element, symbol Rd, atomic number 37, relative atomic mass 85.47. It is one of the alkali metals, ignites spontaneously in air and reacts violently with water. It is used in photoelectric cells and vacuum-tube filament.

Rubidium was discovered spectroscopically by German physicists Robert Bensen and Gustav Kirchhoff in 1861 and named after the red lines in its spectrum.

The discovery of the second "spectral element" occurred in the studies of a rare mineral, lepidolite (called also lilalite because of its lilac colour). For the first time, a detailed chemical analysis of lepidolite was performed by M. Klaproth at the end of the 18th century. But the experienced analyst did not discover alkalis in the mineral. Doubting his own results, Klaproth decided to repeat the analysis and this time (1797) he found the following components: 54.5% silicon dioxide, 38.25% aluminium oxide, 4% potassium oxide and 0.75% manganese oxide. The missing 2.5 percent Klaproth ascribed to the loss of water contained in the mineral. However, no matter what ingenious techniques the chemist tried, he could not determine the content of the two most important components: lithium (it had not been discovered yet by that time) and fluorine; thus, the nature of lepidolite remained obscure.

At the beginning of 1861, a sample of this mineral from Saxony fell into the hands of R. Bunsen and G. Kirchhoff, who separated alkaline components from it and precipitated potassium in the form of chloroplatinate. After a thorough washing, the precipitate was subjected to spectral analysis. On February 23, 1861, the chemists reported the existence of a new alkali metal in lepidolite to the Berlin Academy of Sciences. The scientists asserted that the magnificent dark red colour of the line of the new metal gave them every reason to name the element "rubidium" and assign to it the symbol Rb from the Latin word rubidus, which meant a deep red colour. Then Bunsen and Kirchhoff discovered rubidium in the same mineral spring water in which caesium was found a year before. The rubidium content turned out to be only slightly higher than that of caesium. Metallic rubidium was prepared by R. Bunsen in 1863.

THALLIUM

Thallium (Greek thallos "young green shoot") - soft, bluish-white, malleable, metallic element, symbol Ti, atomic number 81, relative atomic mass 204.37. It is a poor conductor of electricity. Its compounds are poisonous and are used as insecticides and rodent poisons; some are used in the optical-glass and infrared-glass industries and in photoelectric cells.

Discovered spectroscopically in 1861 by its green line, thallium was isolated and named by William Crookes later that year.

Thallium became the third element whose presence in the earth minerals was established by spectroscopy. Some of its properties proved to be similar to those of alkali metals and therefore, there were scientists who believed that thallium was not an independent chemical element but a mixture of alkali metals, namely unknown heavy analogues of rubidium and caesium. Time was required to dispel the doubts. While Bunsen and Kirchhoff continued to investigate the newly discovered elements, their method of spectral analysis attracted attention of the English chemist and physicist W. Crookes. By that time he had been known to the scientific community mainly as the editor and publisher of the Chemical News Journal. There was nothing glamorous in the way Crookes started on his way to the discovery. Back in 1850 he received ten pounds of sludge remaining in lead chambers after production of sulphuric acid in Tilkerod plant (Germany). The scientist separated selenium from the sludge for the study of compounds called selenocyanides to which his first published paper was devoted. After the extraction of selenium and its purification, a certain amount of the material remained and there was every reason to suspect the presence of tellurium, a direct analogue of selenium in terms of chemical properties. However, with the methods he used he could not extract tellurium. The investigation was stopped and it was just a lucky chance that the scientist kept the residue after the processing of the sludge (and perhaps, the belief that the residue contained tellurium).

The discovery of caesium and rubidium impressed W. Crookes very much. Being not only impressionable but practical as well, the scientist understood at once how very promising the spectral method was for analytical purposes. Having obtained a spectroscope, Crookes decided to test it immediately. The time came for the samples of the sulphuric acid sludge (or, to be more exact, its residue after removal of selenium) which had been kept for more than ten years. Crookes introduced the sample into the flame of a burner and was instantly disappointed: no hint of tellurium lines in the spectrum. The selenium lines appeared and then gradually faded. However, instead of them a magnificent green line appeared which Crookes had never observed before. Of course, there was a temptation to assign the line to a new chemical element and the scientist did so naming it "thallium" from the Greek thallos, which means "a new green branch".

The first publication about Crookes' discovery appeared in Chemical News on March 30, 1861 under the title "On the Existence of a New Element probably from the Sulphur Group". Here the author was wrong since, as we know, thallium has nothing in common either with sulphur or with its analogues. A year later Crookes recognized his mistake and published another paper titled "Thallium, a New Chemical Element" where no analogy with sulphur was drawn.

In this way was thallium discovered. The word "discovered" means here the establishing of the existence of thallium by the new method. After having observed the element's spectrum Crookes neither separated the pure element nor prepared its compounds. This was done by the French chemist C. Lamy who is often credited with being an independent discoverer of thallium.

For the first time C. Lamy observed the green thallium line in a sample of selenium extracted from the sludge of sulphuric acid production (the raw material used by Crookes). This took place in March 1862, a year after Crookes' observations and already on June 23 Lamy submitted a sample of metallic thallium with a mass of about 14g to the Paris Academy of Sciences. Crookes also succeeded in preparing metallic thallium but in the form of powder. C. Lamy, however, declared that the thallium of Crookes was nothing other than the metal sulphide. Controversy went on, Crookes said that he had obtained the metal powder before May 1, 1862, but did not dare to fuse the powder into an ingot because of the product's volatility. A special committee organized by the Paris Academy of Sciences, including such prominent scientists as A. Saint Claire Deville, T. Pelouze and J. Dumas, recognized the priority of G. Lamy.

The French chemist undoubtedly studied thallium in much greater detail than W. Crookes. He showed that the metal formed trivalent and monovalent compounds. Monovalent thallium has much in common with alkali metals; trivalent thallium resembles aluminium. J. Dumas named it "the paradoxical metal". It was the similarity of thallium with sodium and potassium that gave rise to the idea that thallium was a mixture of unknown alkali metals with large atomic masses. It is regrettable that all the credit for the discovery of thallium is given to W. Crookes, while the French chemist's significant achievements are often ignored.

In 1866 E. Nordenshold, a well-known traveller, mineralogist and one of the explorers of Greenland, found a new mineral containing silver, copper, selenium and thallium. He proposed to name it Crookesite (in honour of W. Crookes). For a long time this mineral was believed to be the only one containing noticeable amounts of thallium.

INDIUM

Indium (Latin indicum "indigo") - soft, ductile, silver-white, metallic element, symbol in, atomic number 49, relative atomic mass 114.82. It occurs in nature in some zinc ores, is resistant to abrasion and is used as a coating on metal parts. It was discovered in 1863 by German metallurgists Ferdinand Reich (1799-1882) and Hieronymus Richter (1824-1898), who named it after the two indigo lines of its spectrum.

In the history of chemical elements, the discovery of a new element often directly affected the discovery of another one. Thus, the discovery of thallium was a catalyst for the discovery of indium - the last of the classic group of four elements identified by spectral analysis.

The stage was set in the German town of Freiberg and the main characters were F. Reich, professor of physics in the Mining Academy and his assistant Th. Richter. The time was the year of 1863. Interested in some properties of thallium, discovered two years earlier, F. Reich decided to obtain a sufficient amount of the metal for his experiments. Searching for natural sources of thallium, he analysed samples of zinc ores mined at Himmelsfurst. In addition to zinc, the ores were known to contain sulphur, arsenic, lead, silicon, manganese, tin and cadmium, in a word, quite a number of chemical elements. Reich believed that thallium could be added to the list. Although time-consuming, chemical experiments did not produce the desired element, he obtained a strawyellow precipitate of an unknown composition. It was told that when C. Winkler (subsequently the discoverer of germanium) entered Reich's laboratory, the latter showed him a test-tube with the precipitate and said that it contained sulphide of a new element.

It would have been surprising if F. Reich had not used spectroscopy to prove his assumption. Ofcourse, Reich did use it but, unfortunately, he was colour-blind and therefore, asked his assistant Richter to perform spectral analysis.

Th. Richter succeeded in the very first attempt: in the spectrum of the sample he saw an extremely bright blue line which could not be confused with either caesium blue line or any other line. The observation was quite definite. Reich and Richter came to the conclusion that the ores of Himmelsfurst contained a new chemical element. They named it "indium" after "indigo", a bright blue dye. There is an interesting fact that does credit to F. Reich. The first reports about the discovery of indium were signed by the two scientists. Reich, however, believed that this was unjust and that the honour of the discovery belonged solely to Richter.

Soon after the two scientists had proved the existence of natural indium with the help of spectroscopy, they obtained a small amount of it. Indium compounds turn the flame of a Bunsen burner blue-violet and so bright that the presence of the new element could be established without a spectroscope. Subsequently Reich and Richter studied some properties of indium, with Winkler giving them considerable help.

143

When metallic indium, although contaminated, was prepared, Richter submitted the samples to the Paris Academy of Science in 1867 and estimated their value at 800 pounds sterling which was quite a lot of money at the time.

Chemical properties of indium were described soon after its discovery but its atomic mass was at first determined incorrectly (75.6). Mendeleev saw that this atomic mass would not correctly place indium in the periodic table and suggested to increase it by about 50 percent. Mendeleev proved to be right and indium occupied its place in the third group of the periodic table.

7

Rare earths

"It was a sea of errors and the truth was drowning in it," the eminent French chemist G. Urbain once said about the history of rare-earth elements. Although he had a reputation for temperament and expansiveness, in this case he did not exaggerate. Indeed, during thirty-odd years (from 1878 to 1910) more than one hundred discoveries of new rare earths were reported and only ten of them proved to be true. The confused and complicated story of the rare earths is not easy to describe.

Lanthanum $(Z = 57)$ and the following fourteen lanthanides from cerium $(Z - 58)$ to lutetium $(Z = 71)$ are usually classified as rare-earth elements. Two more elements can be added to the list: Yttrium $(Z = 39)$ and Scandium $(Z = 21)$; their properties are similar to those of lanthanum and they are linked historically with the rare earths. It was precisely the discovery of yttrium that began the history of rare-earth elements. Scandium, mentioned only briefly here, is considered in greater detail in chapter 9.

In total, the rare-earth elements (REEs) represent 1/5 of all the natural elements and their discoveries spanned 113 years - from 1794 (the discovery of yttrium) to 1907 (the discovery of lutetium). One of REEs, promethium, was prepared artificially much later. The unusual history of REEs is due to their extraordinary properties and first of all, to their striking chemical similarity. In minerals and ores, they are encountered all together at the same time and it is extremely difficult to break the mixture into constituents. This made the history of REEs very rich in false discoveries with new elements often turning out to be mere combinations of already known ones. Even real discoveries did not always relate to pure rare-earth elements: in many cases the newly discovered elements proved later to be a mixture of two or more unknown elements. That is why the widely accepted dates of the discovery of some REEs must be treated with a pinch of salt.

Another important feature in the history of REEs was that they all were first extracted in the form of oxides. Chemists of the past used the name "earths" for oxides of, for instance, magnesium, calcium ("alkaline earths") and applied it (erroneously, as it became clear later) to oxides of the first REEs, yttrium and cerium. Hence the term "rare earths". Pure metals were prepared long after the discovery of the corresponding elements. For instance, a series of heavy lanthanides were prepared as pure metals only after the Second World War. Therefore, in our subsequent narration, the term REEs will refer to oxides.

Rare-Earth Elements (REEs) Early History

In 1794, the Finnish chemist Johann Gadolin, a chemist at the University of Abo, separated an oxide of an unknown element from ytterbite and named it yttrium. The mineral had been found seven years before in an old quarry at Ytterby, a small Swedish village. The village gave the name to the mineral (although later it was rechristened gadolinite in honour of Gadolin) and then to yttrium and three more REEs: erbium, terbium and ytterbium.

Samples of ytterbite, a mineral, were also studied by other contemporary analysts: L. Vauquelin in France and M. Klaproth in Germany. They also found a new oxide (earth) in it but their values for its content were different. Since the analytical methods were the same, the differences in the results could be explained in the following way: the mineral contained another unknown element whose separation from yttrium was difficult.

It really proved to be so but the "stranger" was found in another mineral. It happened in 1803. J. Berzelius and W. Hisinger on the one hand and M. Klaproth on the other separated an oxide of a new element independently of one another and named it "cerium" after the recently discovered (1801) asteroid Ceres; the mineral was named "cerite". For many years these two minerals, gadolinite and cerite, were the only sources of REEs.

Cerium was very much like yttrium although there were some differences too. It is now known that what was believed to be "cerium" was in effect a complex mixture of cerium REEs (from Ce to Gd) and what was held to be "yttrium" was a mixture of yttrium REEs (from Tb to Lu). Thus, in 1794 and 1803, respectively, no real yttrium and cerium were discovered. In 1826 C. Mosander, a disciple of Berzelius, suspected that cerium extracted from cerite contained an impurity. Thirteen years needed the scientist to turn his conjecture into confidence.

Lanthanum and Didymium, Terbium and Erbium
Lanthanum (Greek lanthanein "to be hidden") - soft, silvery, ductile and malleable, metallic element, symbol La, atomic number 57, relative atomic mass 138.91, the first of the lanthanide series. It is used in making alloys. It was named in 1839 by Swedish chemist Carl Mosander (1797-1858).

Terbium - soft, silver-grey, metallic element of the lanthanide series, symbol Tb, atomic number 65, relative atomic mass 158.925. It occurs in gadolinite and other ores, with yttrium and ytterbium and is used in lasers, semiconductors and television tubes. It was named in 1843 by Swedish chemist Carl Mosander for the town of Ytterby, Sweden, where it was first found.

Erbium - soft, lustrous, greyish, metallic element of the lanthanide series, symbol Er, atomic number 68, relative atomic mass 167.26. It occurs with the element yttrium or as a minute part of various minerals. It was discovered in 1843 by Carl Mosander and named after the town of Ytterby, Sweden near which the lanthanides (rare-earth elements) were first found.

Until C. Mosander began a thorough study of rare earths, yttrium and cerium attracted relatively little attention: they both received the status of chemical elements and their properties were more or less known.

If there had been a tradition of planting a tree in honour of a newly discovered element, yttrium and cerium would have been weak young saplings in this imaginary garden. To continue the analogy, for seventy years, after 1839, these young trees were branching intensively.

After a thorough study of cerium, C. Mosander established that it contained two more new elements lanthanum (La) and didymium (Di). "Lanthanum" originates from the Greek for "to lie hidden" and in fact for a long time lanthanum escaped the attention of researchers. "Didymium" in Greek means a "twin" since it resembles lanthanum as two drops of water resemble each other, and it took C. Mosander's magnificent skill to show that lanthanum and didymium were different elements. The branches on the cerium tree could be illustrated in the following manner:

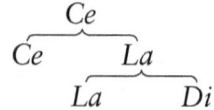

Later, many researchers attempted to encroach on the chemical individuality of cerium and lanthanum. They wanted to prove that these elements were complex. However, C. Mosander obtained relatively pure oxides of these elements. As regards didymium, it had a different fate. You will not find its symbol in the modern periodic table. It is a long story which we shall tell you later. Here we shall only note that the real beginning of cerium biography was the year of 1839. The same is true of yttrium. Mosander began to study yttrium in 1843 inspired by his successful decomposition of cerium. And Gadolin's old yttrium showed its real face. Strictly speaking, there were three faces: yttrium itself and two elements extremely similar to it - terbium and erbium. The situation was as follows:

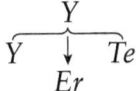

Yttrium asserted its individuality later. Whether Mosander obtained pure terbium or not remains unclear. Erbium had the same fate as didymium. And one more correction in the list of official discovery dates is necessary: real yttrium was extracted by Mosander in 1843. Therefore, it is Mosander who stood at the cradle of REEs.

After Mosander's work, the list of known REEs remained unchanged for almost 40 years. Scientists made a lot of mistakes studying these elements; they gave erroneous formulas of oxides and determined atomic masses incorrectly. Mendeleev was firmly convinced that "something was wrong" and he proposed to change the values of atomic masses of the REEs discovered up to 1869. From the literature on the periodic law, we know that he was absolutely right, but this did not practically affect the further fate of REEs. These elements were so similar in properties that their separation could not be reliably controlled. The situation became paradoxical: a mixture of elements was taken for a single element and vice versa, newly discovered elements proved to be mixtures of elements.

Even spectral analysis, which had played such an important role in the discovery of new elements, yielded results which were more often erroneous than reliable.

"YTTERBIUM", SCANDIUM, "HOLMIUM", THULIUM

Ytterbium - soft, lustrous, silvery, malleable and ductile element of the lanthanide series, symbol Yb, atomic number 70, relative atomic mass 173.04. It occurs with (and resembles) yttrium in gadolinite and other minerals and is used in making steel and other alloys.

In 1878 Swiss chemist Jean-Charles de Marignac gave the name ytterbium (after the Swedish town of Ytterby, near where it was found) to what he believed to be a new element. French chemist Georges Urbain (1872-1938) discovered in 1907 that this was in fact a mixture of two elements: ytterbium and lutetium.

Scandium - silver-white, metallic element of the lanthanide series, symbol Sc, atomic number 21, relative atomic mass 44.956. Its compounds are found widely distributed in nature, but only in minute amounts. The metal has little industrial importance.

Scandium is relatively more abundant in the sun and other stars than on Earth. Scandium oxide (Scandia) is used as a catalyst, in making crucibles and other ceramic parts and scandium sulphate (in very dilute aqueous solution) is used in agriculture to improve seed germination.

The element was discovered and named in 1879 by Swedish chemist Lars Nilson (1840-1899) after Scandinavia, because it was found in the Scandinavian mineral euxenite.

Holmium (Latin Holmia "Stockholm") - silvery, metallic element of the lanthamide series, symbol Ho, atomic number 67, relative atomic mass 164.93. It occurs in combination with other rare-earth metals and in various minerals such as gadolinite. Its compounds are highly magnetic.

The element was discovered in 1878, spectroscopically, by the Swiss chemists L. Soret and Delafontaine and independently in 1879 by Swedish chemist Per Cleve (1840-1905), who named it after Stockholm, near which it was found.

Thulium - soft, silver-white, malleable and ductile, metallic element, of the lanthanide series, symbol Tm, atomic number 69, relative atomic mass 168.94. It is the least abundant of the rare-earth metals and was first found in gadolinite and various other minerals. It is used in arc lighting.

The X-ray-emitting isotope Tm-170 is used in portable X-ray units. Thulium was named by French chemist Paul Lecoq de Boisbaudran in 1886 after the northland Thule.

Almost four decades after Mosander's work the "rare-earth" saplings still did not give any new branches. There were many reasons for this. Scientists could not tackle the capricious chemistry of the REEs. Separation of these elements was based on the fact that their salts differed, although slightly, in solubility. Therefore, to separate one rare earth from another more or less reliably, hundreds of similar recrystallizations had to be performed.

Known rare-earth minerals were few; gadolinite and cerite were extremely rare and the other minerals (there were about ten of them) could be likened, as regards their abundance, to museum pieces. Nevertheless, the era of new discoveries had come and the first sprouts appeared on the yttrium tree. Mosander's erbium remained controversial for a long time and only in 1878 did the Swiss scientist J. De Marignac separate a new element from erbium; he named it "ytterbium" also after the village of Ytterby.

Both in the text and in the heading of this section we put "ytterbium" between quotation marks. This means that ytterbium was not an element properly speaking but, as was shown later, a mixture of some REEs. The names of other newly discovered elements which turned out to be mixtures have also been written in quotation marks. Thus, 1878 cannot be considered to be the final date of the discovery of "ytterbium".

The fact that "ytterbium" was a mixture was established already the following year by the Swedish chemist L. Nilson; he named the discovered element "scandium" in the honour of Scandinavia.

Thus, erbium minus "ytterbium", minus scandium... Could erbium be finally considered to be free of impurities? However, in 1879 Nilson's compatriot P. Cleve showed that erbium without "ytterbium" and scandium was still a mixture; Cleve split it into three components: erbium itself, "holmium" and thulium. "Holmium" was named after the old name of Stockholm and thulium in the honour of the legendary country of Thule at the world's end. And it was no less difficult to isolate thulium than to reach the far and mysterious Thule.

In 1879 the chemical individuality of erbium freed from impurities was proven beyond any doubt and that year rather than 1843 can be considered to be the date of its discovery. Thulium turned out to be pure as well, but "holmium's" real birth was still ahead. So the yttrium tree branched copiously within two years:

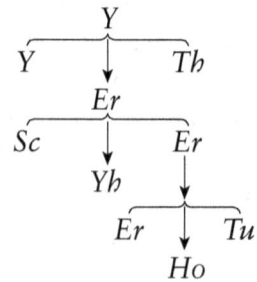

There are certain peaks in the history of elements. The two glorious years 1878-1879 were such a peak in the history of REEs. The period was marked by another important event: deposits of a new rare-earth mineral, samarskite, were found in North America. It is of interest that this name is of a Russian origin. Back in the 1860's, a mineral of a complex composition containing rare earths was found in the Urals. It was named after the mining engineer V.E. Samarskii; the American mineral proved to be identical to the Uralian one.

The importance of this event can hardly be overestimated. The discovery of Samarskite eliminated an acute shortage in rare-earth raw material which became available to many chemical laboratories. When scientists have sufficient amounts of materials to be studied, they can perform more detailed experiments and properly check the results obtained. Samarskite became a producer of new REEs.

And at last, in the late 1870, scientists sufficiently improved the spectroscopic method for it to become a powerful factor in the discovery of new REEs although "production losses" were rather high: the spectra of individual REEs resemble one another just as their chemical properties do.

The End of "Didymium", "Samarium", Neodymium and Praseodymium

Samarium - hard, brittle, grey-white, metallic element of the lanthanide series, symbol Sm, atomic number 62, relative atomic mass 150.4. It is widely distributed in nature and is obtained commercially from the minerals monzanite and bastnasite. It is used only occasionally in industry, mainly as a catalyst in organic reactions. Samarium was discovered by spectroscopic analysis of the mineral samarskite and named in 1879 by French chemist Paul Lecoq de Boisbaudran (1838-1912) after its source.

Neodymium - yellowish metallic element of the lanthanide series, symbol Nd, atomic number 60, relative atomic mass

144.24. Its rose-coloured salts are used in colouring glass and neodymium is used in lasers.

It was named in 1885 by Austrian chemist Carl Von Welsbach (1858-1929), who fractionated it away from didymium (originally thought to be an element but actually a mixture of rare-earth metals consisting largely of neodymium, praesodymium and cerium).

Praseodymium (Greek praseo "leekgreen + dymium") - silver-white, malleable, metallic element of the lanthanide series, symbol Pr, atomic number 59, relative atomic mass 140.907. It occurs in nature in the minerals monzanite and bastnasite and its green salts are used to colour glass and ceramics. It was named in 1885 by Austrian chemist Carl Von Welsbach (1858-1929).

He fractionated it from dydymium (originally thought to be an element but actually a mixture of rare-earth metals consisting largely of neodymium, praseodymium and cerium) and named it for its green salts and spectroscopic line.

"Didymium" is one of the most surprising pages in the history of REEs. Its unprecedented chemical similarity to lanthanum finally convinced scientists that the REE chemistry is a quite special branch of inorganic chemistry. For a long time, the identity of "didymium" was not questioned. Turning over the pages of scientific journals dating to the middle of the last century, we do not find any statements worthy of attention that "didymium" was a mixture of elements.

Mendeleev put the symbol Di into his periodic table and described "didymium" as a separate chemical element although, in general, the great Russian scientist was suspicious about the REEs (for instance, he did not recognize the existence of terbium).

The death sentence to "didymium" was signed by the study of samarskite. At the end of 1878 the French spectroscopist M. Delafontaine began to study didymium extracted from this mineral and found two new lines in its spectrum. Since at that time the accepted approach was "a new line in the spectrum means a new element", Delafontaine thought just that.

In his opinion, a new previously unknown element contained in "didymium" was responsible for the appearance of the new lines in the spectrum. He named it "decipium" from the Latin "to deceive, to stupefy" and the name proved to be ironical: "decipium" turned out to be a mixture of several REEs both known and unknown ones. Decipium was debunked in 1879 by L. de Boisbaudran of France who played a prominent role in the discovery of new REEs. In the next chapter, we shall tell you how he discovered gallium predicted by Mendeleev. Boisbaudran extracted "didymium" from samarskite and thoroughly studied the sample by spectroscopy. Boisbaudran was a much more skilful experimenter than Delafontaine and he succeeded in separating the impurity from "didymium". He named the new element "samarium" after samarskite, being unaware that "samarium" was also a mixture of elements. Boisbaudran's discovery was immediately confirmed by Marignac who, after multiple recrystallizations of "samarium", separated two fractions which he marked Yα and Yß (not to be confused with the symbol of yttrium Y!). The spectrum of the second fraction was identical to the spectrum of "samarium". As to the first fraction, we shall have a look at it a little later.

Thus, indivisible "didymium" gave way to "didymium" and "samarium". Isn't it time to remove the quotation marks from the name "didymium"? Perhaps, having freed itself from "samarium", "didymium" found, at last, its own individuality?

Here a new character in our narration appears - the Czech chemist B. Brauner, a great friend of Mendeleev and an ardent follower of his ideas about periodicity. Beginning with 1875, Brauner persistently studied "didymium" with the sole aim of proving that the element could be oxidized to a pentavalent state. A positive answer would have made it possible to place "didymium" into the fifth group of the periodic table since there was no place for it either in the third or in the fourth group. Besides, the complex problem of placing REEs in the table would have become simpler.

Naturally, Brauner did not obtain pentavalent "didymium". We know now that lanthanides cannot reach this oxidation state. However, trying to determine the atomic mass of "didymium" more correctly, Brauner decided to obtain the element in as pure a form as possible. He discovered that "didymium" separated from samarium could be divided into three fractions somewhat differing in molecular weights. Brauner performed this experiment in 1883 but he had to stop further research for some reasons. It was a great pity since he was so close to ending the story of the old "didymium".

This honour fell to the Austrian chemist C. Auer Von Welsbach who made a great contribution to REEs chemistry. Up to that time rare earths had no practical applications but C. Auer Von Welsbach attracted the attention of engineers to them. At the time the whole world used gas lighting and in 1884 the scientist invented a new incandescent mantle which was impregnated with a special mixture containing REE salts. This sharply increased the brightness of the light and considerably prolonged the service life of the mantles which became to be known as Auer's mantles. Industry demanded hundreds of kilograms of rare-earth minerals. This stimulated the search for new deposits and in 1886 rich deposits of monazite-sand containing large quantities of REEs were found in Brazil. This fully satisfied also the needs of chemists in rare-earth material for studies.

On June 8, 1885, C. Auer Von Welsbach reported to the Viennese Academy of Sciences how he had split "didymium" into two components. He named one of them praseodymium (from the Greek for a "green twin" because of the light green colour of its salts) and the second, neodymium ("new twin").

Not even the name of the old "didymium" survived! The cerium rare-earth "tree" looked now as follows.

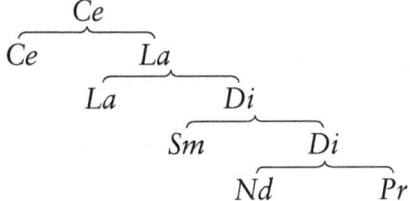

GADOLINIUM AND DYSPROSIUM

Gadolinium - silvery-white metallic element of the lanthanide series, symbol Gd, atomic number 64, relative atomic mass 157.25. It is found in the products of nuclear fission and used in electronic components, alloys and products needing to withstand high temperatures.

Dysprosium (Greek dusprositos "difficult to get near") - silver-white, metallic element of the lanthanide series, symbol Dy, atomic number 66, relative atomic mass 162.50. It is among the most magnetic of all known substances and has great capacity to absorb neutrons.

It was discovered in 1886 by French chemist Paul Lecoq de Boisbaudran (1838-1912).

These two elements complete the history of the REEs in the 19th century; in the case of gadolinium the decisive role was played by G. de Marignac.

We have already mentioned that Marignac succeeded in breaking "samarium" into two fractions: Yα and Yβ. There was no problem with the fraction Yβ but Yα gave a lot of trouble. Marignac was not audacious enough to recognize that this fraction was in effect a new element. This conclusion was made in 1886 by de Boisbaudran. He decided to name the new element gadolinium (in honour of Gadolin, the pioneer of the REEs chemistry) and asked Marignac to give his consent. The consent was received but Marignac's generosity is all the more striking since he had neither claimed co-authorship of the discovery nor put forward any priority claims. However, we believe that the credit for the discovery of gadolinium should go to both scientists.

It is unquestionable that dysprosium was discovered (1886) by de Boisbaudran alone. Having prepared sufficiently pure "holmium", the scientist thoroughly studied its spectrum and discovered two new lines which pointed to the presence of an unknown element. After multiple recrystallizations, he separated the impurity; thus, dysprosium was discovered, as well as holmium. Its name originates from the Greek for "difficult to obtain". The name is symbolic since it is characteristic of the REEs history.

"Time of Confusion" in the History of REEs

If we look at the current list of rare-earth elements, we shall see that almost all of them had been discovered by 1886. Only promethium was unknown (it is quite a peculiar case) and europium and lutetium were to be found in the 20th century. The majority of REEs had already been discovered but who could possibly know about it in the second half of the 1880? Who could state it for certain that the natural treasure-troves of REEs had already been exhausted?

On the contrary, it was much more heartening to think that brilliant discoveries of new rare earths were still ahead and such hope is not easily defeated. In the periodic table, REEs were allotted a large space between barium and tantalum. The difference in their atomic masses was 45 units. A great number of REEs, both known and unknown, could have been squeezed into this space. And nobody could predict how many of them. Twenty, thirty, or forty - any number seemed to be reasonable. And this unsure ground favoured numerous discoveries of new REEs.

Many eminent scientists, who knew the cost of real success, set out enthusiastically to split the known REEs and obtained wonderful results which, after a short period of time, they themselves had to declare erroneous. The discoverer of scandium, L. Nilson and his assistant G. Kruss confidently reported in 1887 that holmium could be divided into four components and dysprosium into three. Seven new REEs were born at once. Brauner, who used to be very cautious about his reports, discovered an impurity in cerium which he named metacerium. And so on.

Scientists relied too much on the spectroscopic method: as soon as a new line was observed in the spectrum, they announced the discovery of a new element. The spectral analysis of that time was relatively young and it was not always possible to establish when the new line really was due to a new element and when it belonged to an impurity of some known element. This was, perhaps, the main cause of false discoveries of REEs. Another was that separation methods were few: only fractional crystallization and fractional precipitation. The first method was based on different solubilities of REE salts and the second one on their different basicities. How could it be established whether the product was a pure earth or contained some impurities? From time to time, the molecular mass of an isolated REE oxide had to be checked. If it remained more or less constant, the aim was attained. This method, however, was too time-consuming and cumbersome.

In the 1880's the periodicity law and the periodic system of Mendeleev were widely recognized. Now any newly discovered element had to be given a place in the periodic table. Almost all REEs remained "homeless" but not because there were no vacant places in the system: there were a lot of them between barium and tantalum but they did not agree with the properties of REEs. If they had been placed among different groups of the table, it would have meant that alien elements had been introduced in all (except for III and IV) groups. That is why Brauner tried so hard to prove that didymium was pentavalent. Since these elements conflicted with the periodic table, it was not difficult to make a lot of mistakes. For the first time in the history of chemical elements, it was suggested that REEs were not elements strictly speaking but varieties of elements, hence the unprecedented similarity in their properties.

The idea belonged to the man whose name we have already come across and shall meet again more than once, namely, the English scientist W. Crookes, the discoverer of thallium. He considered REEs to be modifications of elements and named them meta-elements. Crookes made his conclusions on the basis of spectral investigations but spectral analysis in this case was not equal to the task. P.E. Lecoq de Boisbaudran showed that Crookes' conclusions were erroneous.

That was the end of the meta-element hypothesis. However, even the most fantastic of ideas sometimes contain a grain of truth. Believing ordinary elements to be mixtures of meta-elements, W. Crookes assumed that each element had different varieties of atoms. He even proposed to replace the term "element" with the term "an elemental group".

This assumption of Crookes can be compared with the later ideas that many chemical elements are really mixtures of isotopes. Thus, Crookes anticipated with a surprising accuracy the subsequent concept of the isotopic nature of elements.

We called the end of the 19th century "the time of confusion" in the history of REEs. However, step by step scientists approached the truth. Some of them estimated more or less accurately the possible number of REEs. The Danish physicist H. Thomsen hit the nail on the head: he proposed the number of 15. It was the same Thomsen who suggested the "ladder-like" arrangement of the periodic table still used now. B. Brauner offered to place all REEs into the same group as it is accepted in our days.

Samples of metallic lanthanum, cerium and neodymium were exhibited at the 1900 World Exhibition in Paris as great achievements of science and technology.

YTTERBIUM AND LUTETIUM

G. Urbain, whose name was mentioned in the first lines of this chapter, contributed greatly to the development of REEs chemistry. He perfected the methods of their separation, obtained some oxides in a very pure form (to prepare pure thulium he performed 15000 recrystallizations), redetermined the atomic masses but could not succeed in discovering a new element himself.

Only in 1907 did the scientist have a stroke of luck. Urbain proved that the old "ytterbium" of Marignac was a mixture of two elements. Urbain retained the name for one of them and therefore, the real date of birth of ytterbium is 1907. He named the other element lutetium (in honour of the old name of Paris-Lutetia).

It turned out that when G. Urbain was working with "ytterbium", Von Welsbach (who had debunked didymium) was performing a similar operation. Having splitted "ytterbium", the Austrian chemist consigned this name to oblivion and named the constituents "aldebaranium" and "cassiopeum" borrowing the names from astronomy.

Urbain's article had been published, however, several months earlier thus making him the discoverer of lutetium although in German scientific literature the name "cassiopeum" and the symbol Cp were used for a long time. Many scientists believed that Welsbach's results were more reliable. It was the second case in the history of REEs after cerium when two scientists from different countries claimed priority of discovering a new element. However, there is every reason to add a third name - that of the American chemist C. James. He established independently that "ytterbium" was a mixture of elements but described his experiments after the American scientific community had already become acquainted with the works of Urbain and Welsbach.

Lutetium turned out to be the last natural REE and it ends the rare-earth series. Urbain was, however, of a different opinion. In 1911 he announced the discovery of a new element, celtium, placing it after lutetium in the periodic table. Later it became clear that the finding of celtium was in fact an experimental error. Urbain had interpreted its spectrum incorrectly: the new lines in it were actually due to already known elements.

LESSONS OF REES HISTORY

REEs history is very instructive. It was written by dozens of self-sacrificing and hard-working chemists of several generations and there was no place in it for those who were after easy fame and success. Tedious and endlessly repeated procedures for separating twin-elements required boundless patience.

REEs history is an integral process from which not a single step can be thrown out. The discovery of one element prepared the ground for the discovery of another. Even innumerable errors in the long run were of benefit to the whole process since scientists perfected investigation methods, checking their own results and those of their colleagues. In no other case was a repeated discovery of a new element of such a great value as in the history of REEs. The truth was gradually extricated from a sea of errors.

REEs history was greatly affected by the discovery of new rare-earth minerals. We have already told you about the great importance of discovering deposits of samarskite and monazite which satisfied all the requirements of scientists for materials.

This dependence on material is unparalleled in the history of other elements. And, finally, nothing else posed so many difficulties to the periodic table as the problem of REEs placing: it was not known how many REEs and why their chemical properties were so similar. This similarity was understood only in 1921 when the Danish scientist N. Bohr developed his theory of the periodic system. The physicist succeeded in finding the solution of the problem which evaded chemists for so long. Even in our days the controversy about the best way of placing the REEs in the periodic table is going on.

8

Helium and other inert gases

The six inert gases (presently called inert elements) - helium, neon, argon, krypton, xenon and radon - are extremely scarce in nature. Until recently, the inert gases were considered incapable of forming chemical compounds from which their name of "inert" or "noble" gases (William Ramsay proposed another name, "rare" gases, but it did not find acceptance). Their scarcity and inactivity account for their late discovery, at the very end of the 19th century, when physical methods, particularly spectral analysis and liquefaction of gases, became sufficiently well developed. It is interesting that all inert gases were obtained in a free state (the only state in which they are encountered in nature) within a very short period of time. The decisive role in the discovery of argon, helium, neon, crypton and xenon was played, in fact, by one scientist, W. Ramsay, an outstanding English physicist and chemist who in 1904 received the Nobel Prize in chemistry for this work.

The discoveries of helium and radon stand out as unusual. Radon was discovered as a result of radioactivity studies, or more precisely, owing to the application of the radiometric method. Therefore, we shall deal with it in chapter 11, which is devoted to the history of radioactive elements. The discovery of helium occupies an exceptional place in the history of chemistry. In 1868 a line was detected in the spectra of solar prominences, which could be assigned to none of the elements known on earth. This line was attributed to a new element on the sun which was called "helium". Twenty seven years later helium was first extracted on Earth.

HELIUM

Helium (Greek helios "Sun") - colourless, odourless, gaseous, non-metallic element, symbol He, atomic number 2, relative atomic mass 4.0026. It is grouped with the inert gases, is nonreactive and forms no compounds. It is the second most abundant element (after hydrogen) in the universe and has the lowest boiling (-268.9°c/-452°F) and melting points (-272.2° c / - 458° F) of all the elements. It is present in small quantities in the Earth's atmosphere from gases issuing from radioactive elements (from alpha decay) in the Earth's crust; after hydrogen it is the second lightest element.

Helium is a component of most stars, including the sun, where the nuclear-fusion process converts hydrogen into helium with the production of heat and light. It is obtained by compression and fractionation of naturally occurring gases. It is used for inflating balloons and as a dilutant for oxygen in deep-sea breathing systems. Liquid helium is used extensively in low-temperature physics (cryogenics).

Helium's unusual story attracted attention of many scientists and science historians, but the real sequence of events was distorted in numerous descriptions which overgrew with a lot of fictional details. Even a legend was invented - beautiful and impressive - about the discovery of the sun element but it was far from the truth.

165

The French astronomer J, Janssen and the English astronomer N. Lockyer are considered to be the discoverers of helium. They studied the total solar eclipse of 1868 which was especially convenient to observe on the Indian Ocean shores. In letters sent to the Paris Academy of Sciences and read out at one of its sessions, they wrote that the spectra of the sun photographed during the eclipse contained a new yellow line D_3 corresponding to an unknown element. To commemorate this remarkable event (the discovery of a new element existing on the sun but not on the earth) a special medal was minted.

Everything is wrong in this fascinating story except two dates. First of all, in August 1868, Lockyer was not on the Indian Ocean coast and did not observe the total solar eclipse. Janssen made his observations after the eclipse. They were of great importance for astronomy but not for the history of helium. The French astronomer was the first to observe solar prominences (gigantic ejections of solar matter) not during an eclipse and to describe their nature. Here is the text of the telegram sent by him to the Paris Academy of Sciences: "The eclipse and prominences were observed, the spectrum is remarkable and unexpected; prominences are of a gaseous nature."

Up to that time, scientists had known nothing about the nature of prominences. Now it became clear that they were clouds of gaseous matter and had a complex chemical composition. A detailed description of his observation was given by Janssen in a letter which reached Paris only 40 days later and was two weeks behind the letter of another French astronomer S. Raye. The latter also observed the prominences and made certain conclusions about them. And what was Lockyer doing at the time? Without leaving England, he observed the prominences with the help of a specially designed spectroscope and determined the positions of lines in their spectra. On October 23 he sent a letter to the Paris Academy of Sciences; by a surprising coincidence it was received on the same day as J. Jansseen's letter.

On October 26 the letters of Janseen and Lockyer were read to the session of the Academy but they did not contain a word about either the hypothetical sun element or the line which was later identified as the characteristic line of the helium spectrum. It was only pointed out in the letters that prominences had been observed when the sun was not eclipsed. And the medal was minted precisely to mark this event.

Thus, no helium was discovered on August 18, 1868, either by Janssen or by Lockyer. Their observations provided an impetus for an intensive study of prominences by many astronomers. And only then was it noticed that the spectra of prominences contained a line which could be assigned to none of the elements known on the earth. Most clearly the line was observed by the Italian astronomer A-Secci who later designated it as D_3. Secci's name ought to be placed side by side with those of Janssen and Lockyer. His role in discovering helium was no less than that of his predecessors. Secci, however, assumed that the D_3 line could belong to some known element, for instance, hydrogen, under high pressures and temperatures. If this assumption had not been confirmed, Secci would have agreed to consider D_3 line as corresponding to some element unknown on Earth.

N. Lockyer and E. Frankland tried to solve the problem posed by Secci but they did not notice any changes in the hydrogen spectrum. Therefore, in his article of April 3, 1871, Lockyer already used the expression "a new element X". There are indications that the name "helium" (from the Greek helios for "solar") was proposed by Frankland. The word "helium" was first uttered at a British Association session by its president V. Thomson (Lord Kelvin) on August 3, of the same year. Even if we regard the discovery of helium as "fait accompli", then, it still remained unusual. It was the only element which could not be isolated in a material form. What is helium under ordinary conditions - gas, liquid, or solid? What are its properties? What is its atomic mass and where is its place in the natural series of elements?

None of these questions could be answered even approximately. Besides, Secci's doubt was still not cleared. Thus, a period began in the history of helium when it was only a hypothetical element. There was no consensus on helium. Mendeleev firmly supported Secci's point of view, feeling that the bright yellow line could belong to some other known element at high temperatures and pressures. W. Crookes, however, completely recognized helium's independence and considered it to be a primary matter which gave rise to all other elements via successive transformations.

Sometimes it seemed that helium was not unique in its mysteriousness. Astronomers discovered new lines in the spectra of various cosmic objects: the sun, the stars and nebulae. A number of hypothetical elements appeared, namely coronium, arconium, nebulium, protofluorine. Several years later they were all recognized to be non-existent and only helium survived.

To receive recognition, helium had to show its "earth face" and its "earth" history began with a chance event.

On February 1, 1895, W. Ramsay received a short letter from K. Miers, a British museum employee. By that time Ramsay had already been acclaimed as the discoverer of argon and we may think Miers did not choose him by chance. Miers wrote about the experiments of the American researcher W. Hildebrand, performed at the US Geological Institute as early as 1890. Upon heating of some thorium and uranium minerals (for instance, cleveite) a chemically inactive gas was liberated; its spectrum was similar to that of nitrogen and contained new lines.

Later, Hildebrand himself confessed to Ramsay that he had a temptation to attribute these lines to a new element. However his colleagues were sceptical about the results and Hildebrand stopped his experiments. Miers, however, believed that in the light of numerous cases of nitrogen presence in natural uranates it was reasonable to stage another experiment.

Evidently, Ramsay believed that Hildebrand's inactive gas could be argon; therefore, he agreed with Miers and on February 5, he acquired a small amount of cleveite. Ramsay himself, however, was busy with studying argon and attempting to prepare its compounds and therefore, asked his pupil D. Matthews to carry out the experiment. Matthews treated the mineral with hot sulphuric acid and like Hildebrand, observed the formation of bubbles of a gas resembling nitrogen.

When a sufficient amount of the gas was collected, Ramsay performed its spectral analysis (March 14). The picture was unexpected: the spectrum had a bright band whose lines were not found in the spectra of nitrogen and argon.

Although Ramsay had no sufficient facts to make definitive conclusions, he assumed that cleveite contained, in addition to argon, another unknown gas. Ramsay spent a whole week to obtain this gas in as pure a form as possible. On March 22, he compared the spectra of argon and the unknown gas in the presence of B. Brauner. Ramsay provisionally named this gas "Krypton" from the Greek for "secret", "covered". The name later passed to another inert gas. On March 23 the scientist wrote down in his diary that the bright yellow line of "Krypton" did not belong to sodium and was not observed in the argon spectrum. (In the late sixties it was necessary to prove that the D_3 line of solar helium was not the bright yellow line of sodium; history, as we see, repeated itself).

Not quite sure of his results, Ramsay sent an ampoule with the gas to W. Crookes. A day later a telegram was received from Crookes which read: "Krypton is helium, 587.49; come and see." The figure 587.49 corresponded to the wavelength of the solar helium on a specially calibrated scale. Although these data facilitated the identification of helium on earth, otherwise this discovery was independent.

It became possible for the scientists to comprehensively study helium - a new chemical element which was no longer hypothetical. Helium's complete chemical inactivity was not suspicious: similar inactivity of argon had already been known by that time (1894).

A brief communication about the discovery of helium on earth was first published by Ramsay on March 29, 1895, in the "Chemical News" edited by Crookes. It is interesting that almost simultaneously terrestrial helium was discovered in cleveite by the Swedish Scientist P. Cleve (in whose honour the mineral had been named) and by his assistant A. Lunglet. They, however, were a little too late with their experiments and could only express their disappointment, by no means claiming their priority.

Terrestrial helium received full recognition and no attempts were made to refute Ramsay's results. A little time passed and helium was discovered in other minerals and mineral spring waters. In 1898 helium was found in the earth atmosphere.

ARGON

Argon (Greek argos "idle") - colourless, odourless, non-metallic, gaseous element, symbol Ar, atomic number 18, relative atomic mass 39.948. It is grouped with the inert gases, since it was long believed not to react with other substances, but observations now indicate that it can be made to combine with boron fluoride to form compounds. It constitutes almost 1% of the Earth's atmosphere and was discovered in 1894 by British chemists John Rayleigh (1842-1919) and William Ramsay after all oxygen and nitrogen had been removed chemically from a sample of air. It is used in electric discharge tubes and argon lasers.

If you saw the statement "Inert gases were discovered by H. Cavendish in 1785" you would treat it as a joke. But no matter how paradoxical it seems, it is essentially true. Only the word "discovered" is misused here. One would be equally justified in declaring that hydrogen was discovered by R. Boyle in 1660 or by M.V. Lomonosov in 1745. In his experiments Cavendish only observed "something" whose nature became clear one hundred years later. In one of his laboratory records, Cavendish wrote that passing an electric spark through a mixture of nitrogen with an excess of oxygen, he obtained a small amount of residue, no more than 1/125 the initial volume of the mixture. This mysterious gas bubble remained unchanged under the subsequent action of the electric discharge. It is clear now that it contained a mixture of inert gases, the fact which Cavendish could neither understand nor explain.

The famous English physicist's experiment was described in 1849 by his biographer H. Wilson in the book life of Henry Cavendish. In the early 80's of the 19th century, Ramsay studied the reaction of gaseous nitrogen with hydrogen and oxygen in the presence of a platinum catalyst. Nothing came out of these experiments and Ramsay did not even publish his results. As he recalled later, he had just read the book by Wilson and wrote "Pay attention" against the description of Cavendish's experiment. He even asked the description of Cavendish's experiment. He even asked his assistant C. Williams to repeat the experiment but we do not know the result of the attempt. Most likely, nothing came out of it. The episode, however, turned out to be unforgettable for Ramsay (his "hidden memory", as he called it) and played a certain role in the prehistory of argon's discovery. At first, the English physicist J. Rayleigh was the main character in it and the need for a further development of the atomic and molecular theory was its historic background. It was essential to specify the atomic masses of the elements for the development of the theory. Numerous experiments showed that in the majority of cases the atomic masses were not integers. Meanwhile, as early as 1815-1816 the English physician W. Prout advanced a hypothesis, a landmark in the history of natural sciences, that atoms of all chemical elements consist of hydrogen atoms; thus, atomic masses had to be integers. Therefore, either Prout was wrong or the atomic masses were determined incorrectly.

To remove the discrepancy, new studies of the composition and nature of the gases were required. Rayleigh thought it necessary to determine, first of all, the densities of the main atmospheric gases, nitrogen and oxygen, since their atomic masses could then be calculated on the basis of the density values.

Rayleigh published a short article in the influential English journal Nature on September 29, 1892. It might seem that the article was about a mere trifle; the density of nitrogen separated from atmospheric air differed from that of nitrogen obtained by passing a mixture of air and ammonia over a red-hot copper wire. The difference was very small, only 0.001, but it could not be explained by an experimental error. Atmospheric nitrogen was heavier. Thus, a mystery appeared which was described as "an anomalously high density of atmospheric nitrogen". Nitrogen obtained by any other chemical techniques was always lighter by the same value.

What was the cause of the discrepancy? Ramsay became interested in the problem. On April 19, 1894, he met with Rayleigh and discussed the situation. Each of them, however, remained firm in his previous conviction. Ramsay believed that atmospheric nitrogen contained an admixture of a heavier gas and Rayleigh, on the contrary, felt that an admixture of a lighter gas in "chemical" nitrogen was responsible for the discrepancy.

Rayleigh's view seemed more attractive. The composition of atmosphere had been thoroughly studied for more than a hundred years and it was hardly possible that some components of the air could have remained undetected. It is just the time to remember Cavendish's experiment and for Ramsay's "hidden memory" to work. On April 29, Ramsay sent a letter to his wife in which he wrote that nitrogen, probably, contained some inert gas which had escaped their attention; Williams is combining nitrogen with magnesium and is trying to establish what remains after the reaction. "We can discover a new element."

The letter breathes confidence: an unknown gas is a new element which, like nitrogen, is inactive, i.e. it hardly enters into chemical reactions. To separate the "stranger" from nitrogen, Ramsay tried to bond nitrogen chemically and used the reaction of nitrogen with red-hot-magnesium shavings ($3Mg + N_2 = Mg_3N_2$); this is the only example when chemistry played a role in the discovery of inert gases.

Entering into polemics with himself Ramsay, however, assumed another possibility: the unknown gas is not a new element but an allotropic variety of nitrogen whose molecule consists of three atoms (N_3) like oxygen (O_2 - molecular oxygen and O_3 - Ozone). The absorption of nitrogen with magnesium must be accompanied with the decomposition of the N_2 molecule into atoms; the single N atom could then be added to N_2 forming N_3. Such was Ramsay's thinking and later the assumption about the existence of N_3 became a trump card in the hands of argon's opponents. Fruitless attempts to separate an ozone-like nitrogen continued for more than two months but by the 3rd of August, Ramsay had 100 cm^3 of a gas which was nitrogen with a density of 19.086.

The scientist wrote about his success to Crookes and Rayleigh. He sent an ampoule with the gas to Crookes for spectroscopic investigations; Rayleigh himself collected a small amount of the new gas. In the middle of August, Ramsay and Rayleigh met at a scientific session and made a joint report. They described the spectrum of the gas and underlined its chemical inactivity. Many scientists listened to the report with interest but were surprised: how could it be that air contained a new component? The eminent physicist

O. Lodge even asked: "Didn't you, gentlemen, discover the name of the new gas as well?"

The difficulty about the name was settled in early November when Ramsay suggested to Rayleigh to name it argon (from the Greek for "inactive") taking into account its exceptional chemical inactivity and to assign the symbol A to it (which later became Ar). On November 30, the President of the Royal Society Lord Kelvin (W. Thomson who in 1871 was the first to use the name "helium") publicly described the discovery of a new constituent of the atmosphere as the outstanding scientific event of the year. The nature of the constituent, however, was unclear. Was it a chemical element? Such authorities as D.I. Mendeleev and J. Dewar, the inventor of the flask for storage of liquid air, believed that argon was N_3. The absolute chemical inactivity of argon was a new property previously unknown to chemists and therefore, it was difficult to study the gas (in particular, to determine its atomic mass). In addition, it became clear that argon, unlike all known elemental gases, is monatomic, i.e. its molecule consists of one atom.

At a session of the Russian Chemical Society on March 14, 1895, Mendeleev declared: argon's atomic mass of 40 does not fit the periodic system, hence, argon is condensed nitrogen N_3.

Much time had passed before the many problems presented by the discovery of argon were solved. A certain role was played here by the discovery of helium, which also turned out to be an inert and monatomic gas. The argon-helium pair allowed an assumption to be made that the existence of such gases is a regularity rather than a mere chance and one could expect the discovery of new representatives of this family. However, they were not discovered until three years passed. In the meantime, scientists thoroughly studied the properties of helium and argon, made precise determination of their atomic masses and put forward ideas about the location of both elements in the periodic table.

KRYPTON, NEON AND XENON

Krypton (Greek Kryptos "hidden") - colourless, odourless, gaseous, non-metallic element, symbol Kr, atomic number 36, relative atomic mass 83.80. It is grouped with the inert gases and was long believed not to enter into reactions, but it is now known to combine with fluorine under certain conditions: it remains inert to all other reagents. It is present in very small quantities in the air (about 114 parts per million). It is used chiefly in fluorescent lamps, lasers and gas-filled electronic valves.

Krypton was discovered in 1898 in the residue from liquid air by British chemists William Ramsay and Morris Travers; the name refers to their difficulty in isolating it.

Neon (Greek neon "new") - colourless, odourless, non-metallic, gaseous element, symbol Ne, atomic number 10, relative atomic mass 20.183. It is grouped with the inert gases, is non-reactive and forms no compounds. It occurs in small quantities in the Earth's atmosphere.

Tubes containing neon are used in electric advertising signs, giving off a fiery red glow; it is also used in lasers. Neon was discovered by Scottish chemist William Ramsay and Englishman Morris Travers.

Xenon (Greek Xenos "stranger") - colourless, odourless, gaseous, non-metallic element, symbol Xe, atomic number 54, relative atomic mass 131.30. It is grouped with the inert gases and was long believed not to enter into reactions, but is now known to form some compounds, mostly with fluorine. It is a heavy gas present in very small quantities in the air (about one part in 20 million).

Xenon is used in bubble chambers, light bulbs, vacuum tubes and lasers. It was discovered in 1898 in a residue from liquid air by Scottish chemists William Ramsay and Morris Travers.

A lull began in the history of inert gases. There were several reasons for it; one of them was that scientists were dealing with very small amounts of argon and helium. To isolate them from air, one had to chemically remove oxygen, nitrogen, hydrogen and carbon dioxide. All inert gases constitute a negligible part of the earth's atmosphere but to detect traces of their analogues against the background of argon and helium was an especially difficult problem. Another reason was chemical inactivity of argon and helium. Even the most active reagents (for instance, fluorine) were powerless. Chemists had no way of studying inert gases and only physical methods could bring results. However, better physical methods were required and they were developed during the lull. Scientists developed experimental techniques for analysing small amounts of gases, perfected spectroscopes and devices for determining gas densities. Finally, an event took place that was of extreme importance in the history of inert gases. Two engineers, U. Hampson from England and G. Linde from Germany, invented an effective process for liquefaction of gases. Hampson built an apparatus that produced one litre of liquid air per hour. The success gave an impetus to the creative thought of scientists. In early 1898 M. Travers, Ramsay's assistant, began to design a refrigerating apparatus for preparing large amounts of liquid argon. Since atmospheric gases liquefy at different temperatures, they can easily be separated from one another.

The discoveries of argon and helium are remarkable also in that they set the chemists thinking not only about the nature of chemical inertness (the phenomenon was understood only about a quarter of a century later) but about the periodic law and periodic system which were under a serious threat. Three most important characteristics of argon and helium (atomic masses, zero valence, monatomic molecule) put both gases outside the system. That is why Mendeleev was so readily attracted by the convenient thought about N_3.

History has a striking power of prediction. Argon had not been properly discovered yet, when on May 24, 1894, Ramsay wrote a letter to Rayleigh in which he asked whether it had ever occurred to him that there was indeed a place in the periodic table for gaseous elements. For instance:

Li Be B C N O F   X   X   X
                      Cl
                      Mn Fe Co Ni
                      Br
                      ?   Rd Ru Pd ...

Ramsay assumed that the system's small period could contain a triad of elements similar to those of iron and platinum metals in the great periods. The discoveries of argon and helium gave rise to an idea that these gases could occupy the places of two Xs in Ramsay's graph. The atomic masses of these elements, however (4 and 40, respectively), proved to be too different for He and Ar to be placed in the same period. Gradually, the idea about new triads was relegated to the background and Ramsay proposed to place inert gases at the end of each period. In this case one could even expect the discovery of an element with the atomic mass 20, an intermediate between helium and argon. Ramsay's report at the session of British Association in Toronto in August, 1897, was devoted just to this element. The report was entitled "Undiscovered Gas". Ramsay wanted to describe interesting properties of the gas but thought it unwise not to mention its most remarkable property: the gas had not been discovered yet.

And here again we see the same certainty which permeated Ramsay's letter to his wife on the eve of argon's discovery. But now it was not audacity of a romantic but conviction multiplied by experience. The undiscovered gas turned out to be neon. Owing to a whim of fate (a frequent thing in science) the discovery was preceded by another event. The new gas could, obviously, be discovered by gradual evaporation of liquid air and by analysis of the resulting fractions, the ones lighter than argon being especially interesting. On May 24, 1898, Ramsay and Travers received a Dewar flask with liquid air. Unfortunately (or, rather, fortunately) the amount of air was too small to search for argon's predecessor and the scientists decided to use the material for perfecting the procedure of liquid air fractionation. Having done so, Ramsay and Travers discovered by the end of the day that the fraction that remained was the heaviest one.

For a week the fraction remained neglected until on May 31 Ramsay decided to investigate it. The gas was scrubbed from possible impurities of nitrogen and oxygen and subjected to spectral analysis. Ramsay and Travers were dumbfounded when they saw a bright yellow line which could belong neither to helium nor to sodium. Ramsay wrote down in his diary: "May 31. A new gas. Krypton." Recall that this name was previously given to undiscovered helium. Now the name found its place in the history of inert gases. Krypton, however, was not the gas about which Ramsay made a report. Its density and atomic mass were higher than the predicted ones.

The discovery of neon promptly followed. Ramsay and Travers selected light fractions formed on the distillation of air and discovered a new inert gas in one of them. Ramsay later recollected that the name "neon" (from the Greek neos for "new") had been proposed by Ramsay's twelve-year-old son. In this case the experiment was performed by Travers alone since Ramsay was away. It was on the 7th of June. Then a whole week was required to confirm the results, obtain greater amounts of neon and determine its density. Neon, as had been expected, turned out to be an intermediate between helium and argon although it had not yet been isolated as a pure gas. The problem of complete separation of neon and argon was solved later.

Still another inert gas was to be discovered by Ramsay and Travers. The scientists, however, did not feel as certain as in the case of neon. One day in July, 1898, the colleagues were busy with distilling liquid air and separating it into fractions. By midnight they had collected more than 50 fractions discovering Krypton in the last of them (No. 56). After that upon heating the apparatus one more fraction was collected (No. 57) consisting, mainly, of carbon dioxide traces. Ramsay and Travers argued about the expediency of studying it and at last decided to proceed with the experiment. Next morning the scientists observed the spectrum of fraction No. 57, which turned out to be highly unusual. Ramsay and Travers concluded that it could be attributed to a new gas. Pure xenon, however, was prepared only in the middle of 1900. The name "xenon" originates from the Greek xenos, which means "stranger".

INERT GASES AS FOOD FOR THOUGHT

The discovery of inert gases ranks among the four great scientific events of the end of the 19th century that led to revolutionary changes in natural sciences, the other three being the discovery of X-rays by Roentgen, radioactivity and the electron. This prominence given by scientists to inert gases has many reasons.

The history of their discovery is colourful and exciting. Helium, the mysterious solar element, was discovered on Earth and this fact alone illustrates how inventive and penetrating man's mind became in his striving for deeper and better understanding of nature.

No less mysterious argon sowed confusion among scientists. Its chemical inertness made it impossible to be classified as a chemical element in the ordinary sense of the term since it revealed no chemical properties. There was nothing left for the researchers but to grow accustomed to the idea that there can be elements unable to enter into chemical reactions. The idea proved extremely fruitful. The discovery of inert gases contributed to the development of the zero valence concept. Moreover, forming an independent zero group they added harmony to the periodic system. Almost twenty five years after their discovery, the inert gases helped N. Bohr to develop his theory of the electron shells of atoms. This theory, in its turn, explained the chemical inactivity of the inert gases and their atomic structure became the basis of the concepts of ionic and covalent bonds. Thus, the discovery of inert gases contributed greatly to the development of theoretical chemistry.

In the early 60's, they surprised the scientific community once more. Scientists showed that xenon (mainly) and Krypton can form chemical compounds. Now more than 150 such compounds are known. Such late "debunking" of the myth about the complete chemical inactivity of inert gases is a paradoxical and interesting feature in their history.

Inert gases are among the rarest stable elements on the Earth. Here are the data given by Ramsay: there is one part by volume of helium per 245000 parts of atmospheric air, one of neon per 81000000 and one of argon per 106, one of krypton per 20000000 and one of xenon per 170000000. Since then these figures have remained almost unchanged. Ramsay said that xenon content in air is less than that of gold in sea water. This alone shows how excruciatingly difficult was the discovery of inert gases.

9

Elements predicted from the periodic system

"Without the periodic law, we could not either predict the properties of unknown elements or even determine the lack or absence of some of them. The discovery of elements was a matter of observation alone. Therefore, only blind chance, acumen and foresight led to the discovery of new elements... The periodic law opens a new road in this respect." By these words D.I. Mendeleev expressed the idea that time had come in the history of chemical elements when it had become possible to forecast the existence of elements and to predict their most important properties.

The periodic system served as a basis for this. Even its structure revealed where "blank" spaces remained which had to be filled. Knowing the properties of the already discovered neighbours, one could evaluate the most typical properties of unknown elements and calculate some quantitative parameters (atomic masses, density, melting and boiling points and so on) by means of logical projections and simple arithmetic operations. This required great chemical erudition. Mendeleev possessed such erudition which, in combination with scientific courage and belief in the periodic law, allowed him to make brilliant predictions of the existence properties of several new elements.

Mendeleev's wonderful predictions have long become textbook examples and there is hardly a book on chemistry failing to mention eka-aluminium, eka-boron and eka-silicon, which later were discovered as gallium, scandium and germanium.

This is how the predicted elements compare with the real ones.

The left-hand column gives the properties of eka-aluminium, eka-boron and eka-silicon predicted by Mendeleev; the right-hand column contains modern data about gallium, scandium and germanium. There is no need to comment, so strikingly close are the expected properties to the real ones. Here is how Mendeleev explained the use of the prefix "eka": "In order not to introduce new names for the expected elements, I shall call them by the name of the nearest lowest analogue form."

Eka-aluminium Ea	Gallium Ga
• Atomic mass is about 68	• Atomic mass is 69.72
• The pure element must have low melting point	• Melting point is 29.75°C
• Density of the metal is close to 6.0	• Density is 5.9 (Sol.)
• Atomic volume must be close to 11.5	• Atomic Volume is 11.8
• Does not change in air	• Oxidizes weakly upon heating to redness
• Must decompose water upon boiling	• Decomposes water at high temperature
• Forms alums but not so readily as Al	• Gives alums of the formula $NH_4Ga(SO_4)_2.12H_2O$
• Ea_2O_3 must be readily reduced to metal	• Ga is readily reduced by calcination of Ga_2O_3 in a hydrogen flow
• Ea is more volatile than Al; It will be discovered by spectral analysis	• Ga has been discovered by the spectroscopic method

Eka-boron Eb	Scandium Sc
• Atomic mass is about 44	• Atomic mass is 45.1
• Density is about 3.0	• Density is 3.0
• Atomic volume is about 15	• Atomic volume is 15
• Metal is non-volatile and cannot be discovered by the spectral analysis	• Volatility is low
• Forms basic oxide	• Forms basic oxide
• Must decompose water at elevated temperature	• Decomposes water upon boiling
• Eb_2O_3 is insoluble in water; the density is about 3.5	• Sc_2O_3 is insoluble in water; the density is 3.864
• Eb_2O_3 forms_alums with great difficulty	• Sc_2O_3 forms the double salt $3k_2SO_4.Sc_2(SO_4)_3$
Eka-Silicon Es	**Germanium Ge**
• Atomic mass is about 72	• Atomic mass is 72.60
• Density as about 5.5	• Density is 5.327
• Atomic Volume is about 13	• Atomic volume is 13.57
• Density of EsO_2 is about 4.7	• Density of GeO_2 is 4.280
• Basic properties are weak	• GeO_2 is of an amorphous nature
• $EsCl_4$ will be a liquid with a boiling point of about 90°c	• $GeCl_4$ is a liquid with a boiling point of 83°C
• The ability of Es for deoxidation is low	• Ge reaches the lower oxidation states with difficulty
• There should exist an unstable compound EsH_4	• A readily decomposing GeI_4 is obtained
• There should exist an organometallic compound $Es(C_2H_5)_4$	• $Ge(C_2H_5)_4$ is known

Among odd- or even-numbered elements of the same group adding Sanskrit numerals to the name of the element (eka, dvi, tri, chatur, etc.). (The Sanskrit disappeared long ago but many words in various modern languages originate from it). Thus, the nearest analogue of aluminium in its group is eka-aluminium and so on.

GALLIUM

Gallium - grey metallic element, symbol Ga, atomic number 31, relative atomic mass 69.75. It is liquid at room temperature. Gallium arsenide (GaAs) crystals are used in microelectronics since electrons travel a thousand times faster through them than through silicon. The element was discovered in 1875 by Lecoq de Boisbaudran (1838-1912).

The time of discovery of gallium is known to an hour. "On Friday of August 27, 1875, between 3 p.m. and 4 p.m. I discovered some signs that there can be a new simple body in the by-product of chemical analysis of zinc blende from the Pierfitt mine in the Argele Valley (Pyrenees)." With these words P.E. Lecoq de Boisbaudran began his report to the Paris Academy of Sciences. He described some of the new element's properties and noted that its presence in the ore was ascertained by spectral analysis just as predicted by Mendeleev five years before. Boisbaudran extracted an extremely small amount of the substance and therefore, could not study its properties properly.

On August 29, Boisbaudran suggested to name the element "gallium" after Gaul, the ancient name of France. The scientist continued the investigation of the new element and obtained additional information which he included into his report to the Paris Academy and then sent it to the academic journal. In the middle of November, the journal with the article reached Petersburg where Mendeleev was impatiently waiting for it. There is every reason to believe that Mendeleev had already learnt about gallium though at second hand. Two weeks earlier the Russian Chemical Society had received a report from Paris signed by P. de Clermont. It recounted the discovery of gallium and contained a brief description of its properties. However, it was much more important for Mendeleev to read what the discoverer himself had written.

Mendeleev's reaction was prompt; on November 16, he delivered a report to the Russian Physical Society. According to the minutes of the session, Mendeleev declared that the discovered metal was, most probably, eka-aluminium. Next day he wrote an article in French entitled "Note on the Discovery of Gallium". And finally on Novemeber 18, Mendeleev spoke about gallium at a session of the Russian Chemical Society. Such a spurt of activity is understandable: the great chemist saw an element predicted by him becoming a reality. Mendeleev believed that if further investigation confirmed the similarity of eka-aluminium properties to those of gallium, this would be an instructive demonstration of the periodic law's usefulness.

Six days later (a surprisingly short time), the "Note on the Discovery of Gallium" appeared in the journal of the Paris Academy of Sciences. Boisbaudran's reaction to it is of particular interest. He continued his experiments and prepared the new results for publication. The next article by the French scientist was published on December 6. As before, he complained of the difficulties caused by the extreme scarcity of gallium, described the preparation of the metal by the electrochemical method and discussed some of its properties and suggested that the formula of gallium oxide had to be Ga_2O_3.

Only at the end of the article were there a few words about Mendeleev's note. Boisbaudran admitted that he had read it with great interest since classification of simple substances interested him for a long time. He had never known about Mendeleev's prediction of eka-aluminium properties but it did not matter; Boisbaudran believed that his discovery of gallium was facilitated by his own laws of spectral lines of elements with similar chemical properties. In his opinion, spectral analysis played a decisive role. And not a word that Mendeleev in his prediction of eka-aluminium also underlined the prominent role of spectral analysis in the discovery of the new element. According to

Boisbaudran, Mendeleev's predictions had nothing to do with the discovery of gallium.

However, as Boisbaudran went on studying the properties of metallic gallium and its compounds, his results continued to coincide with Mendeleev's predictions. For instance, in May 1876, the French scientist established that gallium was readily fusible (its melting point is 29.5°C), its appearance remained the same after storage in air and it was slightly oxidized when heated to redness. The same properties of eka-aluminium were predicted by Mendeleev in 1870, who calculated the density of eka-aluminium to be 5.9-6.0 on the basis of the periodic system and the densities of eka-aluminium's neighbours. Lecoq de Boisbaudran, however, making use of his spectral laws, found that the density of eka-aluminium was 4.7 and confirmed the value experimentally. Such a difference (less than two units) might seem small to a layman but it was essential for the future of the periodic law. Up to that time only qualitative characteristics of the predicted properties had been confirmed and density was the first quantitative parameter. And it turned out to be erroneous.

There is a widely known story that Mendeleev, having received Boisbaudran's article citing a low (4.7) density of gallium, wrote him that the gallium obtained by the French chemist was contaminated most likely by sodium used in the process of gallium preparation. Sodium has a very low density (0.98), which could substantially decrease the density of gallium. Hence, it was required to purify gallium thoroughly.

This letter has not been found either in France or in the Mendeleev's archives. There is only indirect evidence from Mendeleev's daughter and the eminent historian of chemistry B. Menshutkin that the letter did exist. However that may be, Mendeleev's views became known to Boisbaudran who decided to repeat the measurements of gallium's density. This time he took into account that Mendeleev's calculations for the hypothetical element's density gave 5.9. And he obtained this value at the beginning of September, 1876. His report about this fact needs no comments. The French scientist became firmly convinced of the extreme importance of the confirmation of Mendeleev's predictions about the density of the new element. Sometime later Lecoq de Boisbaudran sent his photo to the great Russian chemist with the inscription: "With profound respect and an ardent wish to count Mendeleev among my friends. L. De B." Mendeleev wrote under it: "Lecoq de Boisbaudran. Paris. Discovered eka-aluminium in 1875 and named it "gallium", Ga = 69.7."

In autumn 1879, F. Engels became acquainted with a new detailed chemistry textbook by H. Roscoe and C. Shorlemmer. For the first time it contained the story about the prediction of eka-aluminium by Mendeleev and its discovery as gallium. In an article to be later included in his Dialectics of Nature Engels quoted the corresponding text from the book and concluded: "By means of the unconscious application of Hegel's law of the transformation of quantity into quality, Mendeleev achieved a scientific feat which is not too bold to put on a par with that of Leverrier in calculating the orbit of the still unknown planet Neptune."

SCANDIUM

Scandium - silver-white, metallic element of the lanthanide series, symbol Sc, atomic number 21, relative atomic mass 44.956. Its compounds are found widely distributed in nature, but only in minute amounts. The metal has little industrial importance.

Scandium is relatively more abundant in the sun and other stars than on Earth. Scandium oxide (Scandia) is used as a catalyst, in making crucibles and other ceramic parts and scandium sulphate (in very dilute aqueous solution) is used in agriculture to improve seed germination.

The element was discovered and named in 1879 by Swedish chemist Lars Nilson (1840-1899) after Scandinavia, because it was found in the Scandinavian mineral euxenite.

We have already briefly mentioned the discovery of scandium in the chapter devoted to REEs. Although many of scandium's properties are similar to those of rare earths, D. I. Mendeleev predicted that the element would be a boron analogue in the third group of the periodic system. His prediction proved to be accurate enough. Scandium was discovered by the Swedish chemist L. Nilson; on March 12, 1879, his article "On Scandium, a New Rare Metal" was published and on March 24 it was discussed at a session of the Paris Academy of Sciences.

Nilson's results, however, were in many respects erroneous. He considered scandium to be tetravalent and gave, therefore, the formula of its oxide as ScO_2. He did not measure the atomic mass and gave only its probable range (160-180). And finally, Nilson suggested that scandium should be placed in the periodic table between tin and thorium, which ran counter to Mendeleev's prediction.

The discovery of scandium excited the scientific community and Nilson's compatriot P. Cleve set out to study the newly discovered element. He studied it thoroughly for almost five months and came to the conclusion that many results obtained by Nilson were erroneous. Cleve reported to the Paris Academy of Sciences on August 18 and the academicians learnt much new about scandium. It turned out to be trivalent; its oxide's formula was Sc_2O_3; its properties differed somewhat from those determined by Nilson. According to Cleve (and this was especially important) Scandium was the eka-boron predicted by Mendeleev; Cleve showed a table in the left-hand column of which eka-boron properties were given and in the right-hand one those of scandium. The following day Cleve sent a letter to Mendeleev in which he wrote: "I have the honour to inform you that your element, eka-boron, has been obtained. It is scandium discovered by L. Nilson this spring."

And, finally, on September 10 Cleve published a long article about scandium from which it is clear that he had a much better understanding of the new element than Nilson. Therefore, those historians who consider Cleve and Nilson as co-discoverers of scandium are right.

For a long time, Nilson was working under an illusion about some of scandium's properties and refused to recognize its identity with eka-boron. Cleve's investigations, however, impressed Nilson very much; in the long run he was forced to admit that he was wrong, thus doing justice to the prediction power of the periodic system.

All of Mendeleev's predictions were confirmed in the long run. The last to be confirmed was the prediction of the density of metallic scandium; only in 1937 did the German chemist W. Fischer succeed in preparing 98 percent pure scandium. Its density was 3.0 g/cm^3, that is exactly the figure predicted by Mendeleev.

GERMANIUM

Germanium - brittle, grey-white, weakly metallic (metalloid) element, symbol Ge, atomic number 32, relative atomic mass 72.6. It belongs to the silicon group and has chemical and physical properties between those of silicon and tin. Germanium is a semiconductor material and is used in the manufacture of transistors and integrated circuits. The oxide is transparent to infrared radiation and is used in military applications. It was discovered in 1886 by German chemist Clemens Winkler (1838-1904).

In parts of Asia, germanium and plants containing it are used to treat a variety of diseases and it is sold in the west as a food supplement despite fears that it may cause kidney damage.

Among the three elements predicted by Mendeleev, eka-silicon was the last to be discovered and its discovery was to a greater extent than in the case of the two others, due to a chance. Indeed, the discovery of gallium by P. Lecoq de Boisbaudran was directly related to his spectroscopic investigations and the separation of scandium by L. Nilson and P. Cleve was associated with thorough investigation of REEs, which was going on at the time.

Predicting the existence of eka-silicon, Mendeleev assumed that it would be found in minerals containing Ti, Zr, Nb and Ta; he himself was going to analyse some rare minerals in search for the predicted element. Mendeleev, however, was not fated to do it and 15 years had to pass before eka-silicon was discovered.

In summer 1885, a new mineral was found in the Himmelsfurst mine near Freiberg. It was named "argyrodite" since chemical analysis showed the presence of silver, the Latin for which is argentum. The Freiberg Academy of Mining asked the chemist C. Winkler to determine the exact composition of the mineral. Analysis was comparatively easy and soon Winkler found the mineral to contain 74.72% silver, 17.43% Sulphur, 0.66% Iron (II) oxide, 0.22% zinc oxide and 0.31% mercury. But what surprised him was that the percentage of all the elements found in argyrodite added up to only 93.04 percent instead of 100 percent. No matter how many times Winkler repeated the analysis 6.96 percent was missing.

Then Winkler made an assumption that the elusive amount had to be an unknown element. Inspired by the idea, he began to study the mineral carefully and in February 1886 the principal events in the discovery of eka-silicon took place.

On February 6, Winkler reported to the German Chemical Society that he had succeeded in preparing some compounds of the new element and isolating it in a free state. The scientist's report was published and sent to many scientific institutions all over the world. Here is the text received by the Russian Physico-Chemical Society: "The signatory has the honour to inform the Russian Physico-Chemical Society that he found in argyrodite a new non-metal element close in its properties to arsenic and antimony, Which he named "germanium". Argyrodite is a new mineral found by Weisbach in Freiberg and consisting of silver, sulphur and germanium."

Three points in this letter deserve attention: firstly, Winkler considered the new element to be a non-metal; secondly, he assumed its analogy with arsenic and antimony and thirdly, the element had already been named. Originally, Winkler wanted to name it "neptunium" but the name had already been given to another element - a false discovery and the scientist proposed the name "germanium" after "Germany". The name became widely accepted although not immediately.

Later it became clear that germanium is to a great extent amphoteric in nature and hence, Winkler's description of germanium as a non-metal cannot be considered completely erroneous. Much sharper debates revolved around the question the analogue of which element in the system germanium was. In his first report, Winkler suggested arsenic and antimony but the German chemist Richter disagreed with Winkler saying that germanium, most likely, was identical to eka-silicon. Richter's opinion seemed to affect the opinion of the discoverer of germanium and in his letter of February 26 to Mendeleev, Winkler wrote: "At first I thought this element would fill the gap between antimony and bismuth in your remarkable and thoughtfully composed periodic system and that the element would coincide with your eka-antimony, but the facts indicate that here we are dealing with eka-silicon."

Such was Winkler's reply to Mendeleev's letter of congratulation. It is interesting that the antimony-germanium analogy was considered erroneous by Mendeleev but he did not think of germanium as eka-silicon either. Probably, Mendeleev was surprised that the natural source of the new element proved to have nothing in common with that predicted by him earlier (titanium and zirconium ores). The discoverer of the periodic law proposed another hypothesis: germanium is an analogue of cadmium, namely, eka-cadmium.

If the nature of gallium and scandium was established beyond any doubt, as regads germanium, Mendeleev was less certain. This uncertainty, however, soon gave way to certainty and already on March 2 Mendeleev wired to Winkler conceding the identity of germanium and eka-silicon.

Soon an exhaustive article by Winkler entitled "Germanium - a new element" was published in the "Journal of Russian Physico-Chemical Society". It was a new illustration of the brilliant similarity between the predicted properties of eka-silicon and real properties of germanium.

Prediction of Unknown Chemical Elements

The history of gallium, scandium and germanium shows that their discoveries were practically unaffected by the periodic law and periodic system. However, the properties predicted by D.I. Mandeleev for eka-aluminium, eka-boron and eka-silicon coincided with those of gallium, scandium and germanium. Mendeleev had determined the main features of these elements long before they were discovered in nature. Is not this fact a striking evidence of the periodic system's power of prediction?

The discovery of gallium and its identity with eka-aluminium became milestones in the history of the periodic law and in the history of discovery of elements. After 1875, even those scientists who had disregarded the periodic system had to recognize its value. And among them there were top researchers, such as R. Bunsen, the creator of spectral analysis (he once said that to classify elements is the same thing as to search for regularities in the stock exchange quotations) or P. Cleve who had never mentioned the periodic system in his lectures. The discovery of scandium and germanium meant further triumph of Mendeleev's theory of periodicity.

In addition to the classic triad, Mendeleev predicted the existence of other unknown elements. On the whole, as early as 1870, Mendeleev saw about ten vacant places in his table. He saw them, for instance, in the seventh group where there were neither manganese analogues nor a heavy iodine's analogue (the heaviest halogen which had to possess metallic properties).

In Mendeleev's papers we find mention of eka-, dvi- and tri-manganese and of eka-iodine. The scientist firmly believed in their existence. And here we encounter a very interesting fact in the predictions. Eka-manganese (known subsequently as technetium) and eka-iodine (astatine) were synthesized later. Mendeleev, naturally, could not know that they did not exist in nature and firmly believed in their existence since these elements filled in the gaps in the periodic system and made it more logical.

The prediction consists of two stages: prediction of the existence of an element and prediction of its main properties. The first stage was in many respects guess-work for Mendeleev. As yet unknown was the phenomenon of radioactivity making some elements so short-lived that their earthly existence is impossible at all or they exist only because they are products of radioactive transformations of long-lived elements (thorium and uranium).

The second stage was completely within Mendeleev's power and depended on his confidence. Sometimes Mendeleev predicted boldly and resolutely. This was the case with eka-aluminium, eka-boron and eka-silicon: these elements had to be placed in that part of the periodic table where many well-known and well-studied elements had already been located - the region of reliable prediction. Sometimes Mendeleev predicted the properties of unknown elements with extreme caution. Among them were analogues of manganese, iodine and tellurium as well as the missing elements of the beginning of the seventh period: eka-caesium, eka-barium, eka-lanthanum and eka-tantalum. Here Mendeleev was groping in the dark, daring only to estimate atomic masses and suggest formulas of oxides. Mendeleev thought that it was difficult to predict the properties of the unknown elements (including those of REES) whose places were at the boundaries of the system because there were few known elements around them. This was the "grey" area of uncertain prediction. Of course, they included the rare-earth elements.

Finally, in some parts of the periodic table prediction was completely unreliable. They included those mysterious stretches extending in the directions of hypothetical elements lighter than hydrogen and heavier than uranium. Mendeleev never thought that the periodic system had to begin with hydrogen. He even wrote a paper in which he described two elements preceding hydrogen. Only when physicists explained the meaning of the periodic law, his mistake became clear: the nucleus of the hydrogen atom had the smallest charge equal to 1. As regards elements which are heavier than uranium, Mendeleev conceded the existence of a very restricted number of them and never took the liberty of predicting, even approximately, their possible properties. Predictions of this kind did not come until much later when they signaled important events in the history of science.

10

Hafnium and rhenium–two stable elements which were the last to be discovered

The elements with the atomic numbers of 72 and 75 were the last stable elements to be discovered in nature - only in the twenties of this century. They are rare, especially rhenium which is one of the least abundant elements. However, the rareness of hafnium and rhenium is hardly responsible for their late discovery. The reason is the peculiar geochemistry of these elements: they are known as trace elements which do not form ores and minerals in the earth's crust but appear in ores and minerals of other elements as low-concentration impurities. Isomorphism (replacement of ions of some elements in crystal lattices of compounds by those of others when the ionic radii are close) largely accounts for their behaviour. The ionic radii of zirconium and hafnium are almost the same, which is responsible for their chemical similarity (their separation is a difficult problem even now). Hafnium in small amounts often accompanies zirconium and, because of their similarity, is not detected against its background.

Rhenium has no special affinity to minerals of any one of the abundant elements. Therefore, while the existence of hafnium was proved rather easily, rhenium was not discovered definitely until after several years of painstaking search.

Scientists knew what they were looking for, planning beforehand what, where and how they were going to discover: they were after elements No. 72 and No. 75. Hafnium was promptly discovered; as for rhenium, brilliant theoretical predictions at first misfired.

The fates of hafnium and rhenium had something else in common: they were discovered with the help of a new method of spectral analysis (X-ray spectroscopy) consisting in the study of X-ray spectra of elements. In 1914 the English physicist H. Moselev discovered the law which related the wavelength of an element's characteristic X-ray radiation to its number in the periodic system. The law made it possible to predict X-ray spectra. Never before was the discovery of new elements so thoroughly prepared as in the case of hafnium and rhenium.

HAFNIUM

HAFNIUM (Latin Hafnia "Copenhagen") – Silvery, metallic element, symbol Hf, atomic number 72, relative atomic mass 178.49. It occurs in nature in ores of zirconium, the properties of which it resembles. Hafnium absorbs neutrons better than most metals, so it is used in the control rods of nuclear reactors; it is also used for light-bulb filaments.

It was named in 1923 by Dutch physicist Dirk Coster (1889-1950) and Hungarian chemist Georg Von Hevesy after the city of Copenhagen where the element was discovered.

The Institute of Theoretical Physics of the Copenhagen University in Denmark was the birthplace of a new element with z=72; the date of birth was the end of December, 1922, although the article about the discovery appeared in a Scientific Journal only in January, 1923. The Dutch spectroscopist D. Coster and the Hungarian radiochemist G. Hevesy named the element after the ancient name of Copenhagen – Hafnia. N. Bohr, whose role in the discovery of hafnium was decisive, stood at the cradle of the element.

The source of element No. 72 was zircon, a rather common mineral, consisting mainly of zirconium oxide. And it was Bohr who suggested the mineral as a subject of investigation.

Why was the Dutch physicist so confident of success? Let us go back to 1870 when Mendeleev was drawing up his periodic system. He reserved the box under zirconium for an unknown element with the atomic mass about 180. Using Mendeleev's terminology, we could name it eka-zirconium. After Mendeleev's predictions of gallium, scandium, and germanium had come true, the confidence in the existence of eka-zirconium became stronger. The question, however, remained about the properties of this hypothetical element. Mendeleev refrained from definite assessments. Generally speaking, there were two possibilities: either eka-zirconium was part of the IV B-Subgroup of the periodic table, i.e. an analogue of zirconium, or it belonged to the rare-earth family as its heaviest element. Now the time has come to recall the name "Celtium".

Having split ytterbium and separated lutetium, the last of the REEs existing in nature, G. Urbain continued the difficult work of separating heavy rare earths. Finally, he succeeded in collecting the fraction whose optical spectrum contained new lines. This event took place in 1911 but at the time did not attract the attention of the scientific community. Perhaps Urbain himself, having suggested the name for it, was not quite sure that he had really discovered a new element. At any rate, he thought it wise to send samples of celtium to Oxford where Moseley worked. Moseley studied the samples by X-ray spectroscopy but the X-ray photographs turned out to be of a poor quality. Nevertheless, in August 1914, Moseley published a communication in which he firmly stated that celtium was a mixture of known rare earths. The communication remained practically unnoticed. In a word, the discovery of celtium was for a very long time considered to be doubtful, although the symbol Ct sometimes appeared in scientific journals.

Meanwhile N. Bohr was working on the theory of electron shells in atoms which also became the corner-stone of the periodic system theory and, at last, explained the periodic changes in the properties of chemical elements. Bohr also solved the problem which had interested chemists for many years: he found the exact number of rare-earth elements. There had to be fifteen of them from lanthanum to lutetium. Only one REE between neodymium and samarium (later known as promethium) remained unknown. Bohr came to this conclusion on the basis of the laws found by him which governed the formation of electron shells of atoms with increasing Z.

Thus, if celtium were indeed a rare-earth element, Bohr's theory would eliminate it completely. Why couldn't it be eka-zirconium? Having proved that lutetium completed the REE series, Bohr firmly established that element No. 72 had to be a zirconium analogue and could be nothing else. Bohr advised D. Coster and G. Hevesy to look for the missing element in zirconium minerals. Now all this seems to us quite logical and clear but at that time many things were at stake: if element No. 72 could not be proved to be a complete analogue of zirconium, the whole of Bohr's periodic system theory would have been questioned. Having separated hafnium from zirconium, Coster and Hevesy confirmed this theory experimentally just as the discovery of gallium had been a confirmation of Mendeleev's periodic system more than half a century before.

When Urbain read the communication about the discovery of hafnium, he understood that this was the end of celtium. Not everybody can take the bitterness of defeat with dignity.

Urbain was reluctant to part with celtium and continued his attempts to identify it with element No. 72. The French spectroscopist A. Dauvillier came to help; he tried to prove the originality of celtium spectra thus making the "element" one of the rare earths.

Moreover, Urbain and Dauvillier declared that Coster and Hevesy had only rediscovered celtium but nothing much came of it, since hafnium soon came into its own. It was prepared in pure form and new spectral investigations showed that there was nothing in common between hafnium and celtium. What an irony of history! Urbain had everything to be the first to discover hafnium. At the beginning of 1922, he and his colleague C. Boulange analysed thortveitite, a very rare mineral from Madagascar. The mineral contained 8 percent of zirconium oxide and the content of hafnium oxide was even higher. It is the only case when hafnium is contained in the mineral in amounts greater than those of zirconium and, nevertheless, Urbain and Boulange failed to uncover element No. 72. The reason for this lies in the great chemical similarity between zirconium and hafnium.

RHENIUM

RHENIUM (Latin Rhenus "Rhine") – Heavy, silver-white, metallic element, symbol Re, atomic number 75, relative atomic mass 186.2. It has chemical properties similar to those of manganese and a very high melting point (3,1800°C/5,7560°F), which makes it valuable as an ingredient in alloys.

It was identified by Walter Noddack and co-workers in Berlin in 1925 and named after the river Rhine.

As regards history, rhenium had an undoubted advantage over hafnium: nobody had ever questioned the fact that element No. 75 had to be an analogue of manganese, or tri-manganese in Mendeleev's terminology. However, in all other respects there was no certainty.

Let us perform an experiment. If we select at random a few monographs and textbooks where rhenium is discussed we shall see that the authors agree on some things while sharply disagreeing on others. They all agree that rhenium was discovered in 1925 but when it comes to the source from which rhenium was extracted, they disagree. Among minerals mentioned as sources of rhenium are columbite and platinum ore, native platinum and tantalite, niobite and wolframite, alvite and gadolinite. Even an experienced geochemist will be at a difficulty finding his way among so varied a group of minerals.

After these introductory remarks, we may name the discoverers of rhenium: V. Noddack, I. Takke (who later married V. Noddack), and the spectroscopist O. Berg. Their authorship was never contested by anybody. This may be the only case when engineers became interested in the yet undiscovered element. They were aware of the uses of the periodic system. Since tungsten was widely used in electrical engineering, there was every reason to believe that element No. 75 would possess properties even more valuable for this industry. It is highly probable that the first attempts of the Noddacks to find this element were prompted by practical needs.

In 1922, after thorough preparations they set to work. First of all, they collected all reports on the discovery of manganese analogues. Since these discoveries remained unconfirmed, it was tempting to check them. The scientists drew up an extensive program of research: they were going to look for two elements at once since unknown manganese analogues included not only element No. 75 but also its lighter predecessor - element No. 43 with an unusual fate. The periodic table made it possible to predict many of their properties. We can now compare the Noddacks' predictions on rhenium with the actual properties of the element:

Prediction	Modern data
Atomic mass 187-188	186.2
Density 21	20.5
Melting point 3300 K	3323 K
The higher oxide formula X_2O_7	Re_2O_7
Melting point of the higher Oxide 400-500°C	220°C

The agreement is, indeed, excellent. Only the melting point of the oxide proved to be much lower than the expected one whereas on the whole Mendeleev's classical method of prediction was fully confirmed. In other words, the Noddacks had a perfectly good idea about what element No. 75 (and element No. 43) was going to be. Thus, the history of rhenium was closely related to the history of its light analogue.

But where to search for these elements? Predicting the geochemical behaviour of rhenium the Noddacks used to the full the capacity of theoretical geochemistry of that time; they even knew that it had to be a very rare element. They could not know. However, that it was a trace element and that, therefore, what seemed unquestionable to them was in effect open to doubt.

The scientists planned to investigate two groups of minerals: Platinum ores and so-called columbites (tantalites). Four years (from 1921 to 1925) were spent in searching for the wanted elements but in vain. Then a communication appeared about the discovery of hafnium whose existence in nature was proved by X-ray spectroscopy. Undoubtedly, this event gave the Noddacks the idea to use the same method in order to prove the existence of manganese analogues and they turned for help to O. Berg, a specialist in X-ray spectroscopy.

In June 1925, V. Noddack, I. Takke, and O. Berg published an article about the discovery of two missing elements, masurium (No. 43) and rhenium (No. 75) They were found in columbite and in the Uralian platinum and named after two German provinces. The elements' X-ray spectra provided the main confirmation of their existence; but there was no question of extracting the elements and the reasoning of the German scientists was, in general, too involved. However, the article attracted attention and other scientists tried to reproduce the results.

However, no such reproduction followed. A year passed and the Soviet scientist O.E. Zvyagintsev and his colleagues proved irrefutably that the Uralian platinum ore contained no new elements. After that the German Scientists continued to study columbites which varied considerably in composition but, according to the predictions, had to contain mysterious manganese analogues. They subjected the minerals to complex chemical treatment in order to concentrate the unknown elements and performed X-ray spectral analysis. The data obtained were reassuring but definite conclusions would have been premature: the scientists could not obtain any noticeable amounts of elements No. 43 and No. 75 and experimentally determine their properties.

Nobody could reproduce the results obtained by the Noddacks. Their compatriot W. Prandtl even sent his assistant A. Grimm to the Noddacks' laboratory to watch them prepare manganese analogues. Back home, A. Grimm reproduced the entire procedure, perfected it and..., we do not know the extent of his distress about the wasted time. The English Scientist F. Loring and the Czechs Ya. Geirovskii and Y. Druce also doubted the Noddacks' results. Later, Loring, Geirovskii and Druce claimed the priority of discovering element No. 75 by other methods and from other sources. History has retained their names but not as discoverers of rhenium.

The two German scientists believed to have also isolated element No. 43 (known later as technetium). Now we know that they by no means could detect the presence of technetium at the time but, nevertheless, the Noddacks were surer of its discovery than of the discovery of rhenium (the fact which is hardly a feather in their cap). As time passed, the Noddacks became convinced that the range of the minerals for analysis had to be considerably enlarged. The previous geochemical prediction did not, apparently, come true. In the summer of 1926 and in 1927 the Noddacks went to Norway to collect minerals among which were: tantalite, gadolinite, alvite, fergusonite and molybdenite. In the early 1928, the scientists, analysing the minerals, isolated about 120 mg of rhenium mainly from molybdenite (molybdenum sulphide). Earlier it had never been considered as a possible source of manganese analogues.

Thus, rhenium became, at last, a reality. An end was put to doubts and the symbol Re occupied forever box No. 75 in the periodic table; masurium, however, remained an enigma for a long time.

Hence, 1928 is the date of the reliable discovery of rhenium, the final step in the long process of search. As regards the widely accepted date, 1925, it is only a landmark in the prehistory of the element.

Having planned the directions of research, the Noddacks assembled all publications on supposed discoveries of eka-manganese. Their notes were lost during the Second World War but, undoubtedly, the name of the Russian Scientist S.F. Kern and the name of the element "devium" were mentioned in them. This may be the most reliable discovery of a new element of all unreliable discoveries. And it is equally possible that the history of element No. 75 could have begun 50 years earlier.

The events were as follows. In 1877, reports appeared about the discovery of a new metal "devium" named after H. Davy. The reports aroused great interest and Mendeleev suggested inviting S.F. Kern to report to a session of the Russian Chemical Society. The scientists of Bunsen's laboratory in Heidelberg decided to check Kern's results carefully. Later his results were confirmed by two or three other scientists. The most interesting fact was that some chemical reactions proved to be identical to those found later for rhenium. Does it not point to the identity of devium and rhenium?

For some reason or other, S.F. Kern lost interest in his discovery and never returned to the problem after 1878. He had extracted the element from platinum ores, which seems impossible from modern point of view (recall Zvyagintsev's work in 1926). The fact is, however, that platinum ores have a complex and varied composition. The Uralian ore does not contain rhenium but its presence as traces in ores of other deposits has been proven.

S.F. Kern studied a very rare sample of platinum ore from Borneo where by that time the mines had already been abandoned. At the beginning of the 20th Century, the Russian chemist G. Chernik worked on the island. Analysing platinum ores he found a constant mass loss in all samples and tried to explain it by the presence of an unknown element. This element could well be Kern's "devium".

In 1950 Y. Druce devoted a large article to devium. He wrote that if rhenium would be discovered in platinum minerals, this would confirm Kern's discovery. Samples of platinum ores from Borneo can be found now only in a few mineralogical museums of the world. It would be of interest to analyse them thoroughly. This is a case when the history of a chemical element could be partially changed.

11

Radioactive elements

We have already discussed the history of discovery of two natural radioactive elements, that is, uranium and thorium. These elements can fairly easily be found in minerals with chemical analysis since their content is sufficiently high. Other natural radioactive elements (polonium, radon, radium, actinium and protactinium) are among the least abundant elements on Earth. Moreover, they exist in nature only because they are the products of radioactive transformations of uranium and thorium.

These elements belonging to the end of the periodic system could not be determined either with chemical analysis or with spectroscopic techniques. They were present in all the minerals where uranium and thorium were found. But not once did scientists suspect that uranium and thorium contained some impurities. Of course, there were always impurities but their content was too low to shift the weighing pans of a balance or to give rise to a new spectral line.

It was only the discovery of a new physical phenomenon known as radioactivity that presented scientists with a method which contributed to a considerable expansion of our knowledge of the properties and structure of matter and to a significant increase in the number of chemical elements in the periodic system. At the early stage of the studies of radioactivity, three types of radiation were found: alpha rays (fluxes of the nuclei of helium atoms with the positive charge of two), beta rays (fluxes of electrons with the negative charge of one) and gamma rays (these are in fact rays similar to X-rays).

Each radioactive element is described by its half-life, that is, the time required for half the initial amount of a radioactive substance to become disintegrated.

POLONIUM

POLONIUM – Radioactive, metallic element, symbol Po, atomic number 84, relative atomic mass 210. Polonium occurs in nature in small amounts and was isolated from pitchblende. It is the element having the largest number of isotopes (27) and is 5,000 times as radioactive as radium, liberating considerable amounts of heat. It was the first element to have its radioactive properties recognized and investigated.

Polonium was isolated in 1898 from the pitchblende residues analysed by French scientists Pierre and Marie Curie, and named after Marie Curie's native Poland.

Polonium was the first natural radioactive element discovered with the radiometric technique. Back in 1870, the main properties of polonium were predicted by D.I. Mendeleev. He wrote: "Among heavy metals we can expect to find an element similar to tellurium whose atomic weight is greater than that of bismuth. It should possess metallic properties and give rise to an acid whose composition and properties should be similar to those of sulphuric acid and whose oxidizing power is higher than that of telluric acid... The oxide RO_z cannot be expected to have acidic properties which tellurous acid still has. This element will form organometallic compounds but not hydrogen compounds."

Nineteen years had passed and Mendeleev made a significant addition to his description of dvi-tellurium (as he called the unknown element). He predicted the following properties: relative atomic mass 212; forms oxide DtO_3; in a free state the element is a crystalline, low-melting non-volatile metal of a grey colour with a density of 9.8; the metal is easily oxidized to DtO_2; the oxide will have weak acidic and basic properties: a hydride of the element, if it exists at all, must be unstable; the element must form alloys with other metals.

Below readers will see for themselves how accurate were Mendeleev's predictions of the properties of a heavy analogue of tellurium. But these predictions had only an indirect effect on the history of polonium, if any. The discovery of polonium (and then radium) proved to be a significant milestone in the science of radioactivity and gave an impetus to its development.

As one can see from the laboratory log-book of Marie and Pierre Curie, they started to study the Becquerel rays, or uranium rays, on December 16, 1897. First the work was conducted by Marie alone and then Pierre joined her on February 5, 1898. He performed measurements and processed the results. They mainly measured the radiation intensities of various uranium minerals and salts as well as metallic uranium. The results of extensive experiments suggested that uranium compounds had the lowest radioactivity, the metallic uranium exhibited a higher radioactivity and the uranium ore known as pitchblende had the highest radioactivity. These results indicated that pitchblende, probably, contained an element whose activity was much higher than that of uranium.

As early as April 12, 1898 the Curies reported this hypothesis in the proceedings of the Paris Academy of Sciences. On April 14, the Curies started their search for the unknown element with the assistance of the chemist G. Bemont. By the middle of July they finished the analysis of pitchblende. They carefully measured the activity of each product successively isolated from the ore. Their attention was focussed on the fraction containing bismuth salts. The intensity of the rays emitted by this fraction was 400 times that of metallic uranium. If the unknown element really did exist it had to be present in this fraction.

Finally, on July 18 Marie and Pierre Curie delivered a report to a session of the Paris Academy of Sciences entitled "On a new radioactive substance contained in pitchblende". They reported that they had managed to extract from pitchblende a very active sulphur compound of a metal that had previously been unknown. According to its analytical properties it was a neighbour of bismuth. The Curies suggested, if the discovery could be proved, to name the new element in honour of the country where Marie had been born and brought up, that is, polonium after Poland.

The scientists emphasized that the element had been discovered with a new research method (the term "radioactivity", which later became conventional, was first introduced in this report).

The introduction of spectral analysis made it possible to reveal the existence in natural objects of elements that could not be seen, felt or weighed. Now the history repeated itself but the role of indicator was played by radioactive radiation, which could be measured with a radiometric technique. However, the results of the Curies were not faultless. They were wrong in suggesting a chemical similarity between polonium and bismuth. Even a brief look at the periodic system shows that the existence of a heavy analogue of bismuth is hardly possible. But one must not forget that the Curies did not extract pure metal, could not determine its relative atomic mass, and, finally, did not see differences in the spectra of polonium and bismuth. This is why they actually ignored a possible analogy between polonium and tellurium.

Thus, we may regard 18 July, 1898, as the date of just a preliminary discovery of polonium as substantiation of the discovery took quite a long time. The high intensity of radiation from polonium made its study difficult. The radiation was found to consist of only alpha rays with no beta or gamma rays. A strange finding was that the activity of polonium decreased with time and the decrease was rather noticeable; neither thorium nor uranium exhibited such behaviour. This is why some scientists doubted whether polonium existed at all. The sceptics said it was just normal bismuth with traces of radioactive substances.

But in 1902 the German chemist W. Marckwald extracted the bismuth fraction from two tons of uranium ore. He put a bismuth rod into a bismuth chloride solution and observed precipitation of a highly radioactive substance on it which he took for a new element and named radiotellurium. Later he recalled: "I named this substance radiotellurium just for the time being since all its chemical properties suggested placing it into the sixth group into the still unoccupied box for the element with a somewhat higher atomic weight than that of bismuth... The element was more electronegative than bismuth but more electropositive than tellurium; its oxide should also have basic rather than acidic properties. All this corresponded to radiotellurium... The expected atomic weight for this substance was about 210". Later he said that he had got his idea for extracting polonium when analysing the periodic system.

As for the polonium discovered earlier, Marckwald promptly declared it a mixture of several radioactive elements. This led to a stormy discussion of the real nature of polonium and radiotellurium. Most scientists supported the Curies. A later comparison of the two elements revealed their identity. The discovery was credited to the Curies and the name "polonium" was retained.

Though polonium was the first of the new natural radioactive elements, its symbol Po did not appear in the appropriate box in the periodic system. The atomic mass of the element was very difficult to measure. The lines of the polonium spectrum were reliably identified in 1910. It was only in 1912 that the symbol Po occupied its place in the periodic table.

For almost half a century, scientists had to be satisfied to work only with polonium compounds (usually in rather small amounts). The pure metal was prepared only in 1946. High-density layers of metallic polonium prepared by vacuum sublimation have a silvery colour. Polonium is a pliable, low melting metal (melting point 254°C, boiling point 962°C), its density is about 9.3g/cm³. When polonium is heated in air it readily forms a stable oxide; its basic and acidic properties are weakly manifested. Polonium hydride is unstable. Polonium forms organometallic compounds and alloys with many metals (Pb, Hg, Ca, Zn, Na, Pt, Ag, Ni, Be). When we compare Mendeleev's predictions with these properties we see how close they are to the truth.

RADIUM

RADIUM (Latin radius "ray") – White, radioactive, metallic element, symbol Ra, atomic number 88, relative atomic mass 226.02. It is one of the alkaline earth metals, found in nature in pitchblende and other uranium ores. Of the 16 isotopes, the commonest, Ra-226, has a half-life of 1.622 years. The element was discovered and named in 1898 by Pierre and Marie Curie, who were investigating the residues of pitchblende.

Radium decays in successive steps to produce radon (a gas), polonium, and finally a stable isotope of lead. The isotope Ra-223 decays through the uncommon mode of heavy-ion emission, giving off carbon-14 and transmuting directly to lead. Because radium luminesces, it was formerly used in paints that glowed in the dark; when the hazards of radioactivity became known its use was abandoned, but factory and dump sites remain contaminated and many former workers and neighbours contracted fatal cancers.

When the Curies and G. Bemont analysed pitchblende, they noticed a higher radioactivity of one more fraction, apart from the bismuth fraction. After they had succeeded in extracting polonium they started to analyse the second fraction thinking that they could find yet another unknown radioactive element.

The new element was named radium from the Latin radius meaning ray. The birthday of radium was December 26, 1898. When the members of the Paris Academy of Sciences heard a report entitled "On a new highly radioactive substance contained in pitchblende". The authors reported that they had managed to extract from the uranium ore tailings a substance containing a new element whose properties are very similar to those of barium. The amount of radium contained in barium chloride proved to be sufficient for recording its spectrum. This was done by the well-known French spectral analyst E. Demarcay who found a new line in the spectrum of the extracted substance. Thus, two methods - radiometry and spectroscopy - almost simultaneously substantiated the existence of a new radioactive element.

The position of radium among the natural radioactive elements (of course, excluding thorium and uranium) almost immediately proved to be the most favourable one owing to many reasons. The half-life of radium was soon found to be fairly long, namely, 1600 years. The content of radium in the uranium ores was much higher than that of polonium (4300 times higher); this contributed to natural accumulation of radium. Furthermore, the intensity of alpha radiation of radium was sufficiently high to allow an easy monitoring of its behaviour in various chemical procedures. Finally, a distinguishing feature of radium was that it evolved a radioactive gas known as emanation. Radium was a convenient subject for studies owing to a favourable combination of its properties and therefore it became the first radioactive element (again, with the exception of uranium and thorium) to find its permanent place in the periodic system without long delay. Firstly, chemical and spectral studies of radium demonstrated that in all respects it belongs to the subgroup of alkaline earth metals; secondly, its relative atomic mass could be determined accurately enough. To do this sufficient amounts of radium preparation had to be obtained. The Curies worked ceaselessly for 45 months in their ill-equipped laboratory processing uranium ore tailings from Bohemian mines. They performed fractional crystallization about 10000 times and finally obtained a priceless prize - 0.1g of radium chloride. The history of science knows no more noble example of enthusiastic work. This amount was sufficient for measurements and on March 28, 1902, Marie Curie reported that the relative atomic mass of radium was 225.9 (which does not differ much from the current figure of 226.02). This value just suited the suggested position of radium in the periodic system.

The discovery of radium was the best substantiated one among the many alleged discoveries of radioactive elements, which soon followed. Every year more new discoveries were reported. Radium was also the first radioactive element obtained in metallic form.

Marie Curie and her collaborator A. Debierne electrolyzed a solution containing 0.106g of radium chloride. Metallic radium deposited on the mercury cathode forming amalgam. The amalgam was put into an iron vessel and heated under a hydrogen flow to remove mercury. Then grains of silvery whitish metal glistened at the bottom of the vessel.

The discovery of radium was one of the major triumphs of science. The studies of radium contributed to fundamental changes in our knowledge of the properties and structure of matter and gave rise to the concept of atomic energy. Finally, radium was also the first radioactive element to be practically used (for instance, in medicine).

ACTINIUM

ACTINIUM (Greek aktis "ray") – White, radioactive, metallic element, the first of the actinide series, symbol Ac, atomic number 89, relative atomic mass 227; it is a weak emitter of high-energy alpha particles. Actinium occurs with uranium and radium in pitchblende and other ores, and can be synthesized by bombarding radium with neutrons. The longest-lived isotope Ac-227, has a half-life of 21.8 years (all the other isotopes have very short half-lives). Actinium was discovered in 1899 by the French chemist Andre Debierne.

Was it just a chance that polonium and radium were the first to be discovered among radioactive elements? The answer is apparently no. Owing to its long half-life radium can be accumulated in uranium ores. Polonium has a short half-life (138 days) but it emits characteristic high-intensity alpha radiation. Though the discovery of polonium gave rise to a controversy, it soon died off.

The third success of the young science of radioactivity was the discovery of actinium. Soon after they had discovered radium the Curies suggested that uranium ore could contain other, still unknown radioactive elements. They entrusted their collaborator A. Debierne with verification of this idea.

Debierne started his work with a few hundred kilograms of uranium ore extracting the "active principle" from it. After he had extracted uranium, radium and polonium he was left with a small amount of a substance whose activity was much higher than the activity of uranium (approximately, by a factor of 100000). At first, Debierne assumed that this highly radioactive substance was similar to titanium in its chemical properties. Then he corrected himself and suggested a similarity with thorium. Later, in spring of 1899 he announced the discovery of a new element and called it actinium (from the Greek for radiation).

Any textbook, reference book or encyclopedia gives 1899 as the date of the discovery of actinium. But in fact, to say that in 1899 Debierne discovered a new radioactive element - actinium - means to ignore very significant evidence to the contrary.

The real actinium has little in common with thorium but we did not mean this chemical difference as evidence against the discovery of actinium by Debierne. The main argument is as follows. Debierne believed that actinium was alpha active and its activity was 100000 times that of uranium. Now we know that actinium is a mild beta-emitter, that is, it emits beta rays of a fairly low energy which are not that easy to detect. Of course, the primitive radiometric apparatus of Debierne was not capable of doing it.

Then what did Debierne discover? It was a complex mixture of radioactive substances including actinium. But the weak beta radiation of actinium was quite indistinguishable against the background of the alpha rays emitted by the products of actinium decay. It took several years to extract the real actinium from this mixture of radioactive products.

In 1911 the outstanding British radiochemist F. Soddy published a book entitled Chemistry of Radioactive Elements where he described actinium as an almost unknown element. He wrote that its atomic weight was unknown, the mean life-time was also unknown, it did not emit rays (this shows how difficult it was to detect the beta radiation of actinium) and its parent substance was unknown. Much about actinium was still vague.

The evidence presented by Debierne for his discovery of actinium did not seem convincing his contemporaries. It is no wonder that soon another scientist - the German chemist F. Giesel - claimed discovery of a new radioactive element. He also extracted a certain radioactive substance whose properties were similar to those of the rare-earth elements. This fact is closer to the truth in the light of our current knowledge. Giesel named the new element emanium because it evolved a radioactive gas - emanation - which made a zinc sulphide screen to glow. Along with the radiotellurium vs. polonium controversy, there appeared a similar controversy between the supporters of actinium and emanium. The first controversy ended by establishing identity between the elements in question. The second controversy proved to be more complicated and could not be speedily resolved since the behaviour of the third new radioactive element was too wayward. The name of Debierne went into the historical records as the name of the discoverer of actinium. However, the substance extracted by Giesel contained a significant proportion of pure actinium as was shown later. Giesel also succeeded in observing the spectrum of emanium. Many scientists believed that they proved identity of actinium and emanium. Gradually, the controversy lost its edge.

The British radiochemist A. Cameron was the first (1909) to place the symbol Ac into the third group of the periodic system (actually, he was the first to put forward the name radiochemistry for the relevant science). But only in 1913 was the position of actinium in the periodic system established reliably. As increasingly pure actinium preparations were obtained the scientists encountered an amazing situation - the radiation emitted by actinium proved to be so weak that some scientists even doubted if it emits at all. It has even been suggested that actinium undergoes an entirely new, radiationless, transformation. It was only in 1935 that beta rays emitted by actinium were reliably detected. The half-life of actinium was found to be 21.6 years.

For a long time extraction of metallic actinium was just out of question. Indeed, one ton of pitchblende contains only 0.15 mg of actinium while the content of radium is as high as 400 mg. A few milligrams of metallic actinium were obtained only in 1953 after reduction of $AcCl_3$ with potassium vapour.

RADON

RADON – Colourless, odourless, gaseous, radioactive, non-metallic element, symbol Rn, atomic number 86, relative atomic mass 222. It is grouped with the inert gases and was formerly considered nonreactive, but is now known to form some compounds with fluorine of the 20 known isotopes, only three occur in nature; the longest half-life is 3.82 days.

Radon is the densest gas known and occurs in small amounts in spring water, streams and the air, being formed from the natural radioactive decay of radium. Ernest Rutherford discovered the isotope Rn-220 in 1899, and Friedrich Dorn (1848-1916) in 1900; after several other chemists discovered additional isotopes, William Ramsay and R. W. Whytlaw-Gray isolated the element, which they named niton in 1908. The name radon was adopted in 1920.

Radon Rn is the 86th element of the periodic system. It is the heaviest of the noble gases. It is highly radioactive and its natural abundance is so low that it could not be identified when W. Ramsay and M. Travers discovered other inert elements. Only application of the radiometric method made possible the discovery of radon.

What we know as radon at present is the combined name for the three natural isotopes of the element No. 86, which were discovered one by one and called emanations. Their appearance heralded a new stage in the studies of radioactivity as they were the first gaseous radioactive substances.

At the beginning of 1899, E. Rutherford (who lived at the time in Canada) and his collaborator R. Owens studied the activity of thorium compounds. Once Owens accidentally threw open the door to the laboratory where a routine experiment was performed. There was a draught and the experimenters noticed that the intensity of radiation of the thorium preparations suddenly dropped. At first they ignored this event but later they observed that a slight movement of air seemed to remove a larger part of the activity of thorium.

Rutherford and Owens decided that thorium continuously emitted a gaseous radioactive substance, which they called the emanation (from the Latin to flow) of thorium, or thoron.

By way of analogy, it was suggested that other radioactive elements could also involve emanations. In 1900, the German physicist E. Dorn discovered the emanation of radium and three years later Debierne observed the emanation of actinium. Thus, two new radioactive elements were found, namely, radon and actinon. An important observation was that all the three emanations differed only in their half-lives - 51.5s for thoron, 3.8 days for radon, and 3.02s for actinon. The longest-lived element is radon and therefore it was used in all studies of the nature of emanations. All the other properties of emanations were identical. All of them lacked chemical manifestations, that is, they were inert gases (analogues of argon and other noble gases). Later they were found to have different atomic masses. But there was just a single slot for these three elements in the periodic system, immediately below xenon.

Such exclusive situation soon became a rule. Therefore, we shall have to discuss briefly some important events in the history of radioactivity studies. Now we must finish the story of radon. This name remained because radon is the longest lived element among the radioactive inert gases. Ramsay suggested to name it niton (from the Latin for glowing) but this name did not take root.

Radioelements and their families

Before the discovery of polonium and radium, there were seven empty slots in the periodic system between bismuth and uranium. While the number of newly found radioactive elements was small there were no problems with their location in the periodic system. But emanations were a baffling problem. They had identical properties and therefore could not be assigned to different boxes of the periodic system, for instance, to the two empty boxes corresponding to the unknown heavy analogues of iodine and caesium. This would be an unnatural thing to do.

But even if we leave the enigmatical radon family alone, the situation still remains unclear. In 1900 W. Crookes observed a strange phenomenon. After fractional crystallization of a uranium compound, he obtained a filtrate and a precipitate. Uranium remained in the solution but it exhibited no activity. On the contrary, the precipitate did not contain uranium but exhibited a high intensity radioactivity. On the strength of his observations, Crookes made a paradoxical conclusion that uranium was not radioactive by itself, and its radioactivity was due to some admixture which he managed to separate from uranium. As if he had ill premonitions, Crookes refrained from giving the admixture any definite name and referred to it as uranium -x (UX). Later it was found that uranium restores its activity after separation of UX which was just a much more active substance. Thus, UX could be regarded as a new radioactive element.

Two years later E. Rutherford and F. Soddy discovered similar temporary disappearance of activity in thorium. The respective admixture was named (by analogy) thorium-x (ThX). Rutherford and Soddy attempted to find an answer to the fundamental question: what happens with a radioactive element in the process of emission of radiation? Does the chemical nature of the element remain unchanged or does it change? They made a valuable observation that the emanation of thorium was produced by ThX rather than by thorium itself. In other words, they identified the first step of radioactive transformations:

$$Th \rightarrow ThX \rightarrow Em\ Th$$

This was the event that played the decisive part in developing the theory of radioactive decay.

According to Rutherford and Soddy, the mechanism of radioactive decay consists in transformation of chemical elements and in their natural transmutation. This was particularly clear in the case of radium, which converted into radon after emission of alpha radiation. Somewhat later, the alpha particle was found to be a doubly ionized helium atom. The decay of radium gave rise to two new elements, namely, radon and helium:

$$Ra \rightarrow Rn + He$$

This suggestion was soon verified in the experiments of Ramsay and Soddy.

Rutherford and Soddy argued further that all the known radioactive elements were not absolutely independent but were genetically linked to each other (converted successively one into another). These elements can be said to make up three radioactive families - the uranium, thorium and radium families named after the originating element of the respective family. Many questions still remained unanswered. How many radioactive elements make up a family? What elements and the families? And finally, what kind of a "material entity" is a radioactive element and what is its real nature?

The last question was not just an abstract one since starting from the early years of the 20th century, the number of radioactive substances started snowballing and the problem of their arrangement in the periodic system became very acute.

New radioactive substances became known under a variety of names such as radioactive bodies, activities and radioactive elements. Scientists were aware that they encountered new, unknown material entities. Most of them manifested their existence only by their radioactive properties, namely, the radiation intensity, the decay type and the half-life. But nothing or almost nothing could be said about their chemical nature. The old classical chemistry of elements always dealt with weighed quantities of substances so that a new element (or its compound) could be extracted in a material form, its reactions could be studied and its spectrum could be recorded. For most newly discovered radioactive elements all this was unfeasible. Hence, it was not unreasonable to ask whether they were elements in the proper chemical sense of the word.

The first researchers of radioactivity disagreed on this account. The Curies and Debierne assumed that all new radioactive substances were elementary in nature and hence, were new chemical elements. The discoveries of polonium, radium and actinium, apparently, supported this viewpoint and these scientists stubbornly adhered to it even when numerous reports on discoveries of new radioactive substances started to pour in. But this stubbornness only fuelled the controversy.

Rutherford and Soddy held another viewpoint. In their opinion, radioactive substances could have different natures. Proceeding from their concept of radioactive families, they argued that there exist relatively stable radioactive elements, that is, uranium, thorium and radium, which give rise to the families or series of radioactive substances. Their chemical nature is well known and thus, they can be classified as ordinary elements with only the property of radioactivity distinguishing them from other elements. The elements which close the radioactive families are normal stable elements (it was already vaguely surmised that lead had to close the radioactive families). According to Rutherford and Soddy, between these two types of elements there exist intermediate substances whose main feature is instability and which cannot be described in chemical terms. They are not elements in the conventional sense of the word, they are just something like atomic fragments. It was suggested to name them "metabolons" (from the Greek for transforming bodies). This approach did away with the problem of location of these substances in the periodic system.

But the name "metabolon" was not widely accepted. Soddy himself soon came to regarding metabolons as chemically individual substances, just like normal radioactive elements. In 1902 the British physicist G. Martin introduced the term radioelement which will be explained below. Here we shall just emphasize that the terms radioelement and radioactive element are by no means identical though they are sometimes confused in literature.

The entire history of radiochemistry in the first two decades of the 20th century is essentially the search for new radioelements and their genetic links to the earlier discovered ones. The compositions of radioactive families became increasingly clear and the families were acquiring features of systems of radioelements just as the periodic system classified the stable elements. The former radium family proved to be a part of the uranium family but there emerged the new actinium family whose originator could not be identified for a long time (this was definitely done only in 1935). Most radioelements were short-lived products whose half-lives were measured in seconds or, at best, in minutes. It was extremely difficult to determine the chemical nature and the places of radioelements in their radioactive families; even the cumbersome and monotonous work on separation of the rare-earth elements could not be compared to this task, which would need an entire book to describe it. Therefore, we have just to present here the chronological data on the discoveries of radioelements (see tables 1-3).

Table 1 Uranium-238 family

Radioelement	Date of Discovery	Discoverers
Uranium - I	1896*	A. Becquerel
Uranium - X$_1$	1900	W. Crookes
Uranium - X$_2$	1913	K. Fajans, O. Gohring
Uranium - II	1911	H. Geiger, J. Nattal
Ionium	1907	B. Boltwood
Radium	1898	The Curies, G. Bemont
Emanation of radium	1900	E. Dorn
Radium - A	1903	E. Rutherford, H. Barnes
Radium - B	1904	P. Curie, J. Danne
	1903	P. Curie, J. Danne
Radium - C	1903	P. Curie, J. Danne
Radium - C^I	1909	O. Hahn, L. Meitner
Radium - C^{II}	1912	K. Fajans
Radium - D (radiolead)	1900	K. Hofmann, E. Strauss
Radium - E	1904	K. Hofmann, L. Gonder, W. Wolf
	1905	E. Rutherford
Radium - F (polonium)	1898	The Curies
*The date of discovery of uranium radioactivity.		

Ficant contribution to the development of the new science of radiochemistry.

As the composition of radioactive families approached the one we know now, the need for reasonable placement of radioelements in the periodic system became increasingly evident. After all, each of the radioelements manifested chemical similarity to one or another conventional element occupying a certain box in the system. But the number of radioelements was too large. Ramsay described the prevailing situation by the French saying embarrass de richesses (confusing abundance).

By the beginning of the second decade of this century, about 40 radioelements had been discovered. Some groups of elements were so similar in their chemical properties that they could not be separated with any of the available methods. (For instance, all three emanations, then thorium, ionium and radiothorium, and finally, radium and thorium-X).

But the atomic masses of radioelements in each of such groups differed considerably, sometimes by a few units. The situation was indeed confusing. Some scientists suggested leaving many radioelements outside the periodic table, but more creative people were not satisfied with such a solution. In 1909 the Swedish scientists D. Stromholm and T. Svedberg suggested placing several radioelements into one box of the table (soon it was clear that they were right). The British radiochemist A. Cameron supported the idea of the Swedes in 1910.

Though back in 1903 radioactivity was proved to be accompanied with transformation of elements, scientists for a long time could not give a definite answer to the question what exactly happens with a radioelement when it emits the alpha or beta particle. An answer to this question would allow to understand where in the periodic system a given radioelement is shifted

Table 2 Uranium-235 family

Radioelement	Date of Discovery	Discoverers
Uranium - 235 (AcU)	1935	A. Dempster
Uranium - U	1911	G. Antonov
Protactinium	1918	O. Hahn, L. Meitner
	1918	F. Soddy, J. Cranston
Actinium	1899	A. Debierne
Radioactinium	1902	F. Giesel
	1906	O. Hahn

Radioelement	Date of Discovery	Discoverers
Actinium - K	1939	M. Pereil
Actinium - X	1900	A. Debierne
	1904	F. Giesel
	1905	T. Godlewski
Emanation of actinium	1902	F. Giesel
Actinium - A	1911	H. Geiger
Actinium - B	1904	A. Debierne
Actinium - C	1904	H. Brooks
Actinium - C^{I}	1908	O. Hahn, L. Meitner
	1913	E. Marsden, R. Wilson
Actinium - C^{II}	1914	E. Marsden, P. Perkins

Table 3 Thorium-232 family

Radioelement	Date of Discovery	Discoverers
Thorium	1898	H. Schmidt. M. Curie
Mesothorium - I	1907	O. Hahn
Mesothorium - II	1908	O. Hahn
Radiothorium	1905	O. Hahn
Thorium - X	1902	E. Rutherford, F. Soddy
Emanation of thorium	1899	E. Rutherford
Thorium - A	1910	H. Geiger, E. Marsden
Thorium - B	1899	E. Rutherford
Thorium - C	1903	E. Rutherford
Thorium - C^{I}	1909	O. Hahn, L. Meitner
Thorium - C^{II}	1906	O. Hahn

owing to radioactive decay. The structure of an atom was still unknown and any changes in the nature of a radioelement could be identified by comparing its chemical properties to the properties of its product. But this was often extremely difficult to do since radiochemists had to work with exceedingly small amounts of substances. In many instances the chemical "portrait" of a radioelement had to be drawn from the secondary features.

Tenacious work of scientists and accumulation of experimental data made it possible to formulate the law of radioactive displacement. Though many scientists took part in this work, the main contributions were made by F. Soddy and the Polish chemist K. Fajans and therefore this law is known as the Soddy-Fajans law. According to it, alpha decay gives rise to a radioelement displaced two boxes to the left from the starting position in the periodic table while beta decay displaces the product one box to the right. When it was shown that the charge of an atomic nucleus equals the number of the respective element in the periodic system, the above empirical law was explained in the following way: an alpha particle removes from a nucleus a positive charge of two and therefore the number of the starting element (the charge of its nucleus) is decreased by two while emission of a beta particle increases the positive charge of the nucleus by one.

The displacement law provided for harmonious relationship between radioactive families and the periodic system of elements. After several successive alpha and beta decays, the originators of the families converted into stable lead giving rise in the process to the natural radioactive elements found between uranium and bismuth in the periodic table. But then each box in the system had to accommodate several radioelements. They had identical nuclear charges but different masses, that is, they looked as varieties of a given element with identical chemical properties but different masses and radioactive characteristics. In December 1913, Soddy suggested the name isotopes for such varieties of elements (from the Greek for the "common place") because they occupy the same box in the periodic system.

Now it is clear that radioelements are just isotopes of natural radioactive elements. The three emanations are the isotopes of the radioactive element radon, the number 86 in the periodic system. The radioactive families consist of the isotopes of uranium, thorium polonium, and actinium. Later many stable elements were found to have isotopes. An interesting observation may be made here. When a stable element was discovered, this meant simultaneous discovery of all its isotopes. But in the cases of natural radioactive elements individual isotopes were discovered first. The discovery of radioelements was the discovery of isotopes. This was a significant difference between stable and radioactive elements in connection with the search for them in nature. No wonder that the periodic system was badly strained when accommodation had to be found for the multitude of radioelements, - it was a classification of elements, after all, not isotopes. The discovery of the displacement law and isotopy greatly clarified the situation and paved the way for future advances.

PROTACTINIUM

PROTACTINIUM (Latin proto "before"+actinium) – Silver-grey, radioactive, metallic element of the actinide series, symbol Pa, atomic number 91, relative atomic mass 231.036. It occurs in nature in very small quantities, in pitchblende and other uranium ores. It has 14 known isotopes; the longest-lived, Pa-231, has a half-life of 32,480 years.

The element was discovered in 1913 (Pa–234, with a half-life of only 1.2 minutes) as a product of uranium decay. Other isotopes were later found and the name was officially adopted in 1949, although it had been in use since 1918.

The element eka-tantalum predicted by Mendeleev is, perhaps, the only one of the radioactive elements that had been discovered earlier than it is generally recognized. We are talking about the element number 91 situated between thorium and uranium. Its long-lived isotope has a considerable half-life (34300 years) and, therefore, it should be accumulated in the uranium ores; moreover, it emits alpha rays. If we look at the accepted date of its discovery (1918), it would be reasonable to ask why it was discovered so late. We shall answer this question later.

Now let us discuss the family of uranium-238 (see Table 1). The notorious element UX discovered by Crookes, which in fact started the hunt for radioelements, is designated as uranium-x_1, in Table 1. This name was given to it much later, after the discovery of the radioelement designated as uranium-x_2.

In February 1913, Soddy suggested that an unknown radioelement should exist between the element UX of Crookes and the element U-II discovered in 1911 in the uranium family. The properties of the new element, according to Soddy, should be those of eka-tantalum. This hypothetical radioelement seemed to have its rightful place in the fifth group of the periodic system which did not contain any radioelements by a strange whim of nature. Strictly speaking, it was not really strange. Uranium-238 (or U-I), the originator of this family, and U-II, a member of the family, are uranium isotopes; both of them have very long half-lives in comparison with other radioelements. It was not so easy to identify uranium-II against the background of uranium-I. It was just as not easy to detect the precursor of uranium-II, that is, the hypothetical eka-tantalum UX_2.

This was done in mid-March 1913 by K. Fajans and his young assistant O. Göring who detected a new beta-emitting radioelement with a half-life of 1.17 minute and chemical properties similar to those of tantalum. In October of the same year, they clearly stated that UX_2 was a new radioactive element located between thorium and uranium and suggested to name it brevium (from the Greek for "Short-Lived").

The symbol UX_2 took its place in the uranium family but the symbol Bv could hardly be put into box no. 91 of the periodic system though the new element was intensely studied in many laboratories and its discovery was verified by British and German scientists.

At any rate, the statement that element no. 91 was discovered in 1913 does not seem controversial. But why then does not its history start with this date?

If the World War I had not started brevium would, perhaps, have a better fate. But the war put a stop to radiochemical studies and sharply curtailed exchanges of information. Eka-tantalum had to be discovered for the second time.

For a long time the actinium family was the most difficult to understand among the three radioactive families. Which element is its originator? The answer was not clear. If it was actinium then its half-life had to be of the same order as the half-lives of thorium and uranium. This seemed to be unlikely though the half-life defied evaluation. At any rate, it was negligible in comparison with the Earth's age.

Since actinium was regarded as the originator of the family, the question of its precursors was meaningless and this attitude contributed to the delay of the discovery of eka-tantalum. Another suggestion was that the actinium family was not independent but just a branch of the uranium family. This suggestion was discussed by radiochemists back in 1913-1914 by which time brevium had already been discovered. But the discussion yielded no meaningful results and actinium continued to be the head of its family though under false pretences (as almost everybody agreed).

A decisive role in further developments was played by the radioelement UY, a thorium isotope discovered in 1911 by the Russian radiochemist G. Antonov who worked in Rutherford's laboratory. The radioelement UX_1 (also a thorium isotope) in the uranium family emits beta particles and gives rise to brevium (UX_2).

The French scientist A. Picard in 1917 suggested that a similar situation had to prevail at the origin of the family which was still known as the actinium family. His idea, which was confirmed only much later, was that the originator of this family was a third, still unknown uranium isotope (in addition to U-I and U-II). Picard named it actinouranium. When it emits alpha particles it converts into Uy which, in its turn, converts into actinium. An intermediate product of the process should be a radioelement belonging to the fifth group of the periodic system. This sequence of transformations can be written as

$$\text{AcU} \xrightarrow{\lambda} \text{UY} \xrightarrow{\beta} \text{Eka Ta} \xrightarrow{\lambda} \text{Ac}$$

This suggestion simultaneously answered the question about UY whose position in the radioactive family was unclear. This constructive, though fairly bold, suggestion was worth verifying.

In England the next stage in the search for eka-tantalum was carried out by Soddy and his assistant A. Cranston. They were lucky and in December 1917 they wrote a paper on their discovery of eka-tantalum as a product of beta decay of uranium-y. But their data on eka-tantalum was rather poor in comparison with the report by the German chemists O. Hahn and L. Meitner.

The paper by the Germans was published earlier though it was submitted to the journal later than the paper by the British scientists. But the important thing is not the publication data. Hahn and Meitner not only extracted the new radioelement; they conducted all possible studies of its properties, evaluated its half-life and measured the mean free path of alpha particles. The German and British scientists are said to be co-discoverers of element no. 91 though the contribution made by the Germans is, undoubtedly, more significant. The tale of the discovery may be ended with the noble gesture of Fajans who did not claim the discovery of eka-tantalum (though he had every right to do so) but just suggested changing the name brevium to protactinium (from the Greek for "preceding actinium") since the latter radioelement was a much longer-lived isotope.

Thus, the symbol Pa appeared in the periodic system. Its isotope with the longest half-life has a mass number of 231. A few milligrams of pure Pa_2O_5 were extracted in 1927.

FRANCIUM

FRANCIUM – Radioactive metallic element, symbol Fr, atomic number 87, relative atomic mass 223. It is one of the alkali metals and occurs in nature in small amounts as a decay product of actinium. Its longest-lived isotope has a half-life of only 21 minutes. Francium was discovered and named in 1939 by Marguerite Perey to honour her country.

The element no. 87 has a place of its own in the history of radioactive elements. Though its natural abundance is extremely small, it was found originally in nature. But we shall tell its story in detail in the part of the book dealing with artificial elements. This will be better for many reasons.

Here the first part of the book comes to an end.

PART - II

Synthesized Elements

The idea of transmutation (transformation) of elements was born in distant times. The idea was upheld by alchemists for their specific aims. But all attempts to achieve transmutation proved futile. As chemistry was developing into an independent full-fledged science and accumulating knowledge of the structure and properties of matter, the very feasibility of transformation of elements was questioned. By the end of the 19th century, serious scientists ignored this problem though did not dare to refute it definitely.

But at the very end of the century, an event happened which suggested the paradoxical idea that continuous transmutation of elements takes place in nature. This event was the discovery of radioactivity. But only a relatively small part of elements at the very end of the periodic system are subjected to natural transmutation.

Radioactive transformations are independent of human will. All attempts to affect the course of natural radioactive processes failed. When the nuclear model of atomic structure was formulated, it became clear that radioactivity is a nuclear

233

phenomenon. The structural features of nuclei determine the capacity for radioactive decay.

The nuclear charge Z is the primary parameter of a chemical element. When a nucleus emits alpha or beta particles its charge changes so that the nature of the chemical element alters. One element is transformed into another. If we are dealing with a stable chemical element, its nuclear charge Z will never change by itself. It will change if we can restructure its nucleus in some way, decrease or increase the number of protons in the nucleus. Only then will the nuclear charge change and artificial transmutation of a chemical element will take place.

Rutherford was the first to carry out artificial transmutation of elements. In 1919 he bombarded nitrogen with alpha particles and obtained oxygen atoms. This first in history, artificial nuclear reaction, can be described by the following equation:

$$^{14}_{7}N + ^{4}_{z}He \rightarrow ^{17}_{8}O + ^{1}_{1}H$$

or, in a shorter form:

$$^{14}_{7}N(\alpha,P)\,^{17}_{8}O$$

Alpha particles for a long time remained the only available means for conducting nuclear reactions. The energy of naturally produced alpha particles is not high; therefore, they could penetrate the nuclei of only a relatively small number of elements and such events were extremely rare. This limited the scope of artificial transmutation of elements. The situation changed significantly as a result of two discoveries made in the thirties. In 1932 the British scientist J. Chadwick discovered a neutral elementary particle known as neutron. Being electrically neutral, neutron proved to be a universal instrument for performing nuclear transformations since it was not repulsed by positively charged nuclei. Two years later, the French physicists

Irene and Frederic Joliot Curie discovered artificial radioactivity and detected a new type of radioactive transformation, namely, positron decay, that is, emission of positrons. It became clear that radioactive isotopes could be produced artificially by means of nuclear reactions for many stable elements.

One can ask what made possible the production of artificial radioactive isotopes in large numbers. The answer is that it was the work of experimental physicists who designed fine instruments for conducting measurements, developed special techniques for performing and studying nuclear reactions and, together with chemists, found methods for isolating traces of radioactive substances. Moreover, the range of particles available for bombardment of nuclei was extended considerably when alpha particles, protons and neutrons were joined by deutrons (nuclei of a heavy hydrogen isotope deuterium), and later by multiply charged ions of such elements as boron, carbon, nitrogen, oxygen, neon, etc. Finally, physicists have built powerful accelerators capable of accelerating charged particles to very high velocities. All these advances paved the way for artificial synthesis of new elements.

12

Discoveries of synthesized elements within the old boundaries of the periodic system

This chapter could be headed "Synthesis of the Missing Elements of the Periodic System". After the discovery of the last stable element, rhenium, only four elements (Nos. 43, 61, 85 and 87) were missing in the table between hydrogen and uranium. All of them were synthesized before the World War II (or purposeful attempts to synthesize them were made). At any rate, the history of synthesized elements starts with the work on these four elements.

TECHNETIUM

TECHNETIUM (Greek technetos "artificial") – Silver-grey, radioactive, metallic element, symbol Tc, atomic number 43, relative atomic mass 98.906. It occurs in nature only in extremely minute amounts, produced as a fission product from uranium in pitchblende and other uranium ores. Its longest-lived isotope, Tc-99, has a half-life of 216,000 years It is a superconductor and is used as a hardener in steel alloys and as a medical tracer.

It was synthesized in 1937 (named in 1947) by Italian physicists Carlo Perrier and Emilio Segre, who bombarded molybdenum with deuterons, looking to fill a missing spot in the periodic table of the elements (at that time it was considered not to occur in nature). It was later isolated in large amounts from the fission-product debris of uranium fuel in nuclear reactors.

The upper part of the periodic system down to the sixth period (where the rare-earth elements are located) always seemed relatively quiet, particularly after the discovery of the group of noble gases which harmoniously closed the right-hand side of the system. It was quiet in the sense that one could hardly expect any sensational discoveries there. The debates concerned only a possible existence of elements that were lighter than hydrogen and elements lying between hydrogen and helium. On the whole, we can say in the parlance of mathematicians that this part of the periodic system was an ordered set of chemical elements.

Therefore, the more awkward and confusing seemed to be the mysterious blank slot no. 43 in the fifth period and seventh group.

Mendeleev named this element eka-manganese and tried to predict its main properties. A few times the element seemed to have been discovered but soon it proved to be an error. This was the case with ilmenium allegedly discovered by the Russian chemist R. Hermann back in 1846. For some time even Mendeleev tended to believe that ilmenium was eka-manganese. Some scientists suggested placing devium (see the end of chapter 10) between molybdenum and ruthenium. The German chemist A. Rang even put the symbol Dv into this box of the periodic table. In 1896 there flashed and burned like a meteor lucium supposedly discovered by P. Barriere.

Mendeleev did not live to see the happy moment when eka-manganese was really found. A year after his death, in 1908, the Japanese scientist M. Ogawa reported that he had found the long awaited element in the rare mineral molybdenite and named it nipponium (in honour of the ancient name of Japan). Alas, Asia once more failed to contribute a new element to the periodic system. Ogawa, most probably, dealt with hafnium (which was also discovered later).

Chemists grew accustomed to a few chemical elements being discovered every year and they were at a loss in the case of eka-manganese. They began to think that Mendeleev could make a mistake and no manganese analogues existed.

H. Moseley decisively refuted this scepticism in 1913. He clearly demonstrated that these analogues have their own place among the elements. In a paper dated September 5, 1925, W. Noddack, I. Tacke, O. Berg announced that they had discovered, together with element no. 75 (rhenium), its lighter analogue in the seventh group of the periodic system, namely, masurium whose number was 43. Two new symbols, Ma and Re, appeared in the periodic table, in chemical textbooks, and numerous scientific publications. The discoverers saw nothing odd in the fact that masurium and rhenium had not been discovered earlier. These elements were thought to be not too rare, however. The lateness of their discovery was attributed to another cause. A large group of trace elements is known to geochemistry. The trace elements are classified as those elements which have no or almost no own minerals but are spread in various amounts over minerals of other elements as if the nature has sprayed them with a giant atomizer. This is why the traces of masurium and rhenium were so hard to identify. Only the powerful eye of X-ray spectral analysis could distinguish them against the formidable background of other elements. There is an ancient saying that if two people do the same things this does not mean that the results will be identical. Two biographies started under the same conditions typically follow different paths. The same can be said about the fates of elements 43 and 75; one of them went a long way and found its proper place while the other's way soon led it to a forest of errors, misunderstandings, and controversies. This was the path of masurium.

W. Prandtl got interested in the empty slots in the seventh group of the periodic table. He had his own outlook and put forward original ideas on the structure of the periodic system. He did not compile a new version of the table, though. He suggested placing the rare-earth elements each to a group though by that time most chemists had put down such an arrangement. But in Prandtl's version of the table, the seventh group happens to reveal as many as four empty slots below manganese corresponding to yet undiscovered elements (this was in 1924) whose numbers were 43, 61, 75 and 93. Prandtl believed this to be no chance occurrence but a result of a common cause that had prevented four elements from having been discovered. The German scientist, however, made his table structure too elaborate and artificial to be accepted. The final discovery of rhenium was the first indication of his errors, and his ideas on the first transuranium element (No. 93) were little thought of at the time. But he was intuitively right in thinking of a close common link between elements 43 and 61.

The belief in masurium's existence gradually diminished. Only the original discoverers were firm. As late as the beginning of the thirties I. Noddack continued to say that in time element 43 would be commercially available as it happened with rhenium. But as time passed and chemists again and again failed to find masurium in whatever minerals they analysed, they came to believe that I. Noddack was right only by half, that is, only about rhenium. Rarest mineral specimens were tested for masurium. Some people even went as far as to claim that masurium minerals were yet to be found and would possess unheard-of properties. Naturally, geochemists were quite sceptical. The imagination of some people went even further and masurium was suggested to be radioactive. That was too much, others said. But it was precisely this shot that did not go wild.

Let us talk about some concepts of nuclear physics. We have discussed isotopes. Now we meet another term - isobars- elements having the same atomic weights or mass numbers but different atomic numbers (from the Greek for "heavy"). Isobars, in other words, are isotopes of different chemical elements with different nuclear charges but identical mass numbers. Take, for instance, potassium-40 and argon-40 which have different nuclear charges (respectively, 19 and 20). Their mass numbers are identical because their nuclei contain different number of protons and neutrons but their total numbers are the same; potassium nucleus contains 19 protons and 21 neutrons while argon nucleus has 20 protons and 20 neutrons.

Thus, the concept of isobars turned out to be the magic key that opened the door to the mystery of masurium.

When the majority of stable chemical elements were found to have isotopes - up to ten isotopes per element - the scientists started to study the laws of isotopism. The German theoretical physicist J. Mattauch formulated one of such laws at the beginning of the thirties (the basic premise of this law was noted back in 1924 by the Soviet chemist S. Shchukarev). The law states that if the difference between the nuclear charges of two isobars is unity, one of them must be radioactive. For instance, in the ^{49}K-^{40}Ar isobar pair the first is naturally weakly radioactive and transforms into the second owing to the so-called process of K-capture.

Then Mattauch compared with each other the mass numbers of the isotopes of the neighbours of masurium, that is, molybdenum (z=42) and ruthenium (z=44):

Mo isotopes	94	95	96	97	98	–	100	–	–
Ru isotopes	–	–	96	–	98	99	100	101	102

What did he deduce from this comparison? The fact that the wide range of mass numbers from 94 to 102 was forbidden for the isotopes of element 43 or, in other words that no stable masurium isotopes could exist.

If that was really so that meant a peculiar anomaly linked to the number 43 in the periodic system. All the atom species with Z = 43 had to be radioactive as if this number was a small island of instability amidst a sea of stable elements. This, of course, would be unfeasible to predict within the framework of purely chemical theory. When Mendeleev predicted his eka-manganese he could never imagine that this member of the seventh group of the periodic system could not exist on Earth.

Of course, in those times (the early thirties) Mattauch's law was no more than a hypothesis, though one that looked like quite capable of becoming a law. And it became just that. The physicist's idea opened the eyes of chemists who had lost all hope of finding element 43 and they saw the source of their errors. However, the symbol Ma remained in box 43 of the periodic system for a few more years. And not without a reason. All right, all masurium isotopes are radio-active. But we know radioactive isotopes existing on Earth - uranium-238, thorium-232, potassium-40. They are still found on Earth because their half-lives are very long. Masurium isotopes are, perhaps, long-lived, too? If so, one should not be too hasty in dismissing the chances of successful search for element 43 in nature.

The old problem remained open. Who knows which way the biography of masurim would take if not for the dawn of a new age - that of artificial synthesis of elements.

Nuclear synthesis became feasible after invention of the cyclotron and the discoveries of neutrons and artificial radioactivity. In early thirties, a few artificial radioisotopes of known elements were synthesized. Syntheses of heavier-than-uranium elements were even reported. But physicists just did not dare to take the challenge of the empty boxes at the very heart of the periodic system. It was explained by a variety of reasons but the major one was enormous technical complexity of nuclear synthesis. A chance helped. At the end of 1936, the young Italian physicist E. Segre went for a post-graduate work at Berkley (USA) where one of the first cyclotrons in the world was successfully put into operation. A small component was instrumental in cyclotron operation. It directed a beam of charged accelerated particles to a target. Absorption of a part of the beam led to intense heating of the component so that it had to be made from a refractory material, for instance, molybdenum.

The charged particles absorbed by molybdenum gave rise to nuclear reactions in it and molybdenum nuclei could be transformed into nuclei of other elements. Molybdenum is a neighbour of element 43 in the periodic system. A beam of accelerated deutrons could, in principle, produce masurium nuclei from molybdenum nuclei.

That was just Segre's thought. He was a competent radiochemist and understood that if masurium really were produced its amount would be literally negligible and its separation from molybdenum would present an enormously intricate task. Therefore, he took an irradiated molybdenum specimen with him back to the University of Palermo where he was assisted in his work by the chemist C. Perrier.

They had to work for nearly half a year before they could present their tentative conclusions in a short letter to the London journal Nature. Briefly, the letter reported the first in history artificial synthesis of a new chemical element. This was element 43 the futile search for which on Earth wasted so much efforts of scientists from many countries. Professor

E. Lawrence from the University of California at Berkley gave the authors a molybdenum plate irradiated with deutrons in the Berkley cyclotron. The plate exhibited a high radioactivity level which could hardly be due to any single substance. The half-life was such that the substances could not be radioactive isotopes of zirconium, niobium, molybdenum and ruthenium. Most probably they were isotopes of element 43.

Though the chemical properties of this element were practically unknown Segre and Perrier attempted to analyse them radiochemically. The element proved to be closely similar to rhenium and exhibited the same analytical reactions as rhenium. However, it could be separated from rhenium with the technique used for separating molybdenum and rhenium.

The letter was written in Palermo and dated June 13, 1937. It was by no means a sensation. The scientific community regarded it as just the authors going on record. The reported data were too patchy while what was needed to be convincing was precisely the details and clear results of radiochemical analysis.

Only later Segre and Perrier were recognized as heroes; indeed, they extracted from the irradiated molybdenum just 10^{-10}g of the new element - an amount formerly undetectable. Never before radiochemists had worked with such negligible amounts of material. The discoverers suggested naming the new element technetium from the Greek for "artificial". Thus, the name of the first synthesized element reflected its origin. The name, though, became generally accepted only ten years later.

Perrier and Segre received new specimens of irradiated molybdenum and continued their studies. Their discovery was confirmed by other scientists. By 1939, it was understood that bombardment of molybdenum with deutrons or neutrons produces at least five technetium isotopes. Half-lives of some of them were sufficiently long to make possible substantial chemical studies of the new element. It no longer sounded fantastic to speak about "the chemistry of element 43". But all attempts to measure accurately the half-lives of the technetium isotopes failed. The available estimates were disheartening since none of them exceeded 90 days and this put a stop to all hopes of finding the element of Earth.

So what was technetium in the late thirties and early forties? Nothing more than an expensive toy for curious scientists. Any prospects of accumulating it in a noticeable amount were, apparently, non-existent. The fate of technetium (and not only of it) was reversed when nuclear physics discovered an amazing phenomenon - fission of uranium by slow neutrons.

When a slow neutron hits a nucleus of uranium-235 it in effect breaks the nucleus down into two fragments. Each of the fragments is a nucleus of an element from the central part of the periodic table, including technetium isotopes. It is not without a reason that a fission reactor (a large-scale nuclear energy producer) is known as a factory of isotopes.

Cyclotron made possible the first ever synthesis of technetium and fission reactor allowed the chemists to produce kilograms of technetium. But even before the first fission reactor started operating, Segre in 1940 found the technetium isotope with a mass number of 99 in uranium fission products in his laboratory. Having found its new birthplace in a fission reactor, technetium started to turn into an everyday (paradoxical as it may be) element. Indeed, fission of 1g of uranium-235 gives rise to 26mg of technetium-99.

As soon as technetium ceased to be a rare bird, scientists found the answers to many questions that had puzzled them, and first of all about its half-lives. In the early fifties, it became clear that three of technetium isotopes are exceptionally long-lived in comparison with not only its other isotopes but also many other natural isotopes of radioactive elements. The half-life of technetium-99 is 212000 year, that of technetium-98 is one and a half million years, while that of technetium-97 is even more namely, 2600000 years. The half-lives are long but not long enough for primary technetium to be conserved on Earth since its origin. The primary technetium would survive on Earth if its half-life were not shorter than one hundred fifty million years. This makes obvious the hopelessness of all search for technetium on Earth.

But technetium can still be produced in the course of natural nuclear reactions, for instance, when molybdenum is bombarded by neutrons. How can free neutrons appear on Earth? They can be produced in spontaneous fission of uranium. The process occurs as described above, only spontaneously, and gives rise to a few neutrons, apart from two large fragments, i.e. nuclei of lighter elements.

The search for technetium in molybdenum ores failed and scientists turned their attention to another possibility. If technetium isotopes are produced in fission reactors why cannot they be born in natural processes of spontaneous uranium fission?

Using as a basis the Earth uranium resources (taking the figure for the mean abundance of uranium in the 20 km thickness of the Earth crust) and assuming the same proportion of produced technetium as in the case of reactor fission, we can calculate that there is just 1.5 kg of technetium on Earth. Such a small amount (though it is larger than for other synthesized elements) could hardly be taken seriously. Nevertheless, scientists attempted to extract natural technetium from uranium minerals. This was done in 1961 by the American chemists B. Kenna and P. Kuroda. Thus, technetium acquired another birthday - the day when it was discovered in nature. If the methods of artificial synthesis of technetium had failed to materialize, even then it would, sooner or later, be brought to light from the bowels of the Earth.

But ten years earlier, in 1951, sensational news about element 43 was heard. The American astronomer S. Moore found characteristic lines of technetium in the solar spectrum. The spectrum of technetium had been recorded immediately when it had become feasible, that is, when a sufficient amount of the element had been synthesized. The spectral data had been compared with those reported by the Noddacks and Berg for masurium. The spectra had proved to be quite different making ultimately clear the mistake of the discoverers of masurium. The spectrum of the solar technetium was identical to that of the terrestrial technetium. An analogy with helium was apparent - both elements sent messages from the Sun before to be found on Earth. True, some astronomers questioned the data on the solar technetium. But in 1952 the cosmic technetium once more sent a message when the British astrophysicist P. Merril found technetium lines in the spectra of two stars with the poetic names of R Andromedae and Mira Ceti. The intensities of these lines evidenced that the content of technetium in these stars was close to that of its neighbours in the periodic system, namely, niobium, zirconium, molybdenum, ruthenium, rhodium and palladium. But these elements are stable while technetium is radioactive. Though its half-life is relatively long it is still negligible on cosmic scale. Therefore, the existence of technetium on stars can mean only that it is still born there in various nuclear reactions. Chemical elements continue to be produced in stars on a gigantic scale. A witty astrophysicist named technetium the acid test of cosmogonic theories. Any theory of the origin of elements must elucidate the sequence of nuclear reactions in stars giving rise to technetium.

PROMETHIUM

PROMETHIUM – Radioactive, metallic element of the lanthanide series, symbol Pm, atomic number 61, relative atomic mass 145. It occurs in nature only in minute amounts, produced as a fission product/by-product of uranium in pitchblende and other uranium ores; for a long time it was considered not to occur in nature. The longest-lived isotope has a half-life of slightly more than 20 years.

Promethium is synthesized by neutron bombardment of neodymium, and is a product of the fission of uranium, thorium or plutonium. It can be isolated in large amounts from the fission-product debris of uranium fuel in nuclear reactors. It is used in phosphorescent paints and as an X-ray source.

The history of one rare-earth element is so unusual that it merits individual discussion. Promethium, as it is known now, is practically non-existent in nature (we write practically but non-absolutely and the reason for that will be clear later). The event which can only be described as amazing preceded the discovery of element 61 by means of nuclear synthesis.

The work of Moseley made clear the existence of an unknown element between neodymium and samarium. But the situation proved to be not so clear and dramatic events rapidly followed in the history of element 61.

The New World was unlucky in discoveries of new elements. All the elements known by the twenties of this century (not counting the elements known from ancient times) had in fact been discovered by the European scientists. This is why the American scientific community was particularly happy to learn about the discovery of element 61 by the chemists from Chicago B. Hopkins, L. Intema and J. Harris in 1926.

Starting from 1913, scientists from various countries had been searching intensely for the elusive rare-earth element and it seemed strange that they had not found it earlier. Indeed, the elements of the first half of the rare-earth family known as the cerium elements (from lanthanum to gadolinium) had been shown by geochemists to be more abundant in nature than the yttrium elements of the second half of the family (from terbium to lutecium). But all the yttrium elements had been found while an empty box had remained in the cerium group between neodymium and samarium.

The straightforward explanation was that element 61 was not just rare but rarest element. Its abundance was assumed to be much lower than that of other rare earth elements, and the available analytical techniques were not sensitive enough to identify its traces in the terrestrial minerals. New more sensitive methods were needed for the purpose.

The American chemists employed X-ray and optical spectral techniques to study the minerals where they hoped to find element 61. Those well versed in the history of rare-earth elements could say that the path the Americans took was a troublesome one as spectral analysis not infrequently had acted as an evil genius of rare earth studies despite all the benefits it had brought to them. But in the twenties the feet spectroscopy stood on were not so unsteady as a few decades earlier and the Moseley law could be used for predicting the X-ray spectra of any element.

The American chemists worked hard, analysed numerous specimens of various minerals and in April 1926 reported the discovery of element 61. But they did not extract even a grain of the new element and its existence was inferred from the X-ray and optical spectral data.

The discoverers named the element illinium in honour of the Illinois University where they worked and the symbol Il took its place in box 61 of the periodic system. But just a half-year later, a new claimant of box 61 came into the lime-light. It had been discovered by two Italian scientists, L. Rolla and L. Fernandes who had named it florencium (Fl). Allegedly, they had discovered element 61 two years earlier than the Americans but failed to report the discovery owing to some undisclosed reasons. They had sealed the report of their discovery into an envelope and left it for safe- keeping in the Florence Academy.

If different people obtain the same result with different means that would seem to prove that the result is genuine. Americans and Italians could be only too happy. As for the question of priority it was nothing new to science. But no one of the alleged discoverers of element 61 could imagine that their argument about priority would soon become superfluous and both symbols, II and F1, would be shown to be illegal squatters in box 61 of the periodic table.

To trace the events now we have to go not further but some time back to the facts that were simply unknown at the time. The report of the discoverers of element 61 started with the words: "There had been absolutely no grounds for assuming the existence of an element between neodymium and samarium until it was demonstrated through the Moseley law." Typical dry style of a scientific report, everything would seem to be correct. But...

The following remarkable conclusion in German (please, do not look it up in a dictionary yet) appeared in the margin of a hand-written manuscript of the element table found in the papers of a certain scientist (we shall supply the name a little later): "NB 61 ist das von mir 1902 vorhergesagte fehlende Element."

The real history of element 61 should prominently feature the name we have already met on these pages. It is the Czech scientist Boguslav Brauner, Mendeleev's friend and an eminent expert in the chemistry of rare-earth elements.

Illinium had been discovered, the discoverers accept congratulations and learn about the second, third, fourth confirmation of the discovery from the scientists of other countries. The pedigree of element 61 could be started thus: "Moseley had predicted and American chemists discovered." But a discordant note unexpectedly sounded in November 1926 from the pages of Nature. It was none other than Brauner. He congratulated his American colleagues but voiced his disagreement with the above-cited beginning of their report. He argued that it was really not important who first discovered element 61 - Americans or Italians; in the twenties scientists became increasingly aware that the discovery by itself was a purely technical matter. The important issue is who predicted the new element. Was it Moseley? No, declared the Czech scientist. Who then? Of course, he himself, Boguslav Brauner...

But nothing could be further from the truth if we thought that he was immodest. His claim was based on his vast experience of work with rare earths, on his profound understanding of the spirit of the periodic system, on his superb appreciation of slight changes of properties in the series of extremely similar rare-earth elements, and finally on his intuition of a dedicated researcher.

But these words of praise must be substantiated with facts. Let us turn back to 1882. The old didymium of K. Mosander is close to its death. P. Lecoq de Boisbaudran has already extracted a new element, samarium, from it. B. Brauner carefully analyses the residue and employing extremely complicated chemical procedures separates it into three fractions with different atomic masses. Owing to a number of reasons he has to discontinue his work and in 1885 K. Auer Von Welsbach overtakes the Czech scientist. The old didymium is dead but praseodymium and neodymium have appeared, the first and the third fractions of Brauner. But what about the intermediate second fraction? No, its time has not come. The chemistry of rare-earth elements is in a turmoil. The muddy stream of erroneous discoveries of new elements overflows with doubts the very periodic system. But life goes on. The chaos in rare earths gradually diminishes and the known rare-earth elements form an ordered series. Now Brauner notices that the difference between with atomic masses of neodymium and samarium is rather large; it is larger than the respective difference between any two neighbouring rare earth elements. His brilliant knowledge of rare earths suggests to Brauner that there is a discontinuity in the variations of their properties in the part of the series between neodymium and samarium. At last, he recalls his work of 1882. The clues fit into a pattern leading to a premonition and even certainty that an unknown element can be found between neodymium and samarium. But as his friend, Mendeleev, Brauner was never too hasty in his conclusions. It was only in 1901 that he placed an empty box between neodymium and samarium when he put forward his views on the place of the rare-earth elements in the periodic system.

Now we can give a translation of the note he wrote in margin of his hand-written table of elements. It reads: "61st element is the missing element predicted by me in 1902."

His short letter to Nature was an attempt by Brauner to put the record straight. This would seem to simplify the task of science historians in writing the history of element 61. But a history is meaningful only if it treats a subject which really exists. As for illinium the element proved to be still born.

While the hotheads kept trying to squeeze the symbol II into box 61 of the periodic table, meticulous critics tried to verify the discovery. The careful experiments by the first of them, Prandtl, could be doubted by nobody. But his results did not even hint at the existence of element 61.

In 1926 the Noddacks who had just announced their discovery of masurium and rhenium (Nos. 43 and 75) started their tests. They used all available techniques to analyse fifteen various minerals suspected of containing illinium. They processed 100 kilograms of rare-earth materials and could not detect a new element. The Noddacks claimed that if the Americans' results had been correct they, the Noddacks, would undoubtedly extracted the new element. Even if the element were 10 million times rarer than niodymium or samarium they would still find it... There are two possible explanations: either element 61 is so rare that the existing experimental techniques are not fine enough to find it or wrong mineral specimens were taken.

Geochemists were against the first explanation. The abundances of rare earth elements are more or less similar. There are no reasons to think that illinium is an exception.

They suggested looking for illinium in minerals of calcium and strontium. All rare-earth elements are typically trivalent but some of them can exhibit a valence for two or four. For instance, europium rather easily gives rise to cantions with a charge of two. Their size is closer to those of calcium and strontium cations and they can replace the latter in the respective alkaline earth minerals. Perhaps, illinium has a similar more pronounced capacity and can be found in some rare natural compound of strontium. One hypothesis replaced another, one assumption stemmed from another, unsubstantiated one. Just in case, the Noddacks analysed several alkaline earth minerals. Alas, they failed once more.

The search for illinium seemed to come to a dead end; though it still went on the reported results were little believed.

Chemists failed in looking for element 61 in the terrestrial minerals. It was theoretical physics whose fate it was to open up the "envelope" where nature had "sealed" element 61. But when the envelope was opened, the scientists (not for the first time) were disappointed. The envelope was empty.

At this point the fate of element 61 directly involved the fate of element 43, that is, technetium. According to the law formulated by the German theoretical physicist Mattauch, technetium in principle cannot have stable isotopes. This law also forbids existence of stable isotopes of element 61. Illinium is dead but element 61 must survive.

But what if it really does not exist? I. Noddack put forward a daring idea that illinium (we shall use this name for the time being) had existed on Earth in early geological periods. But it had been a highly radioactive element with a short half-life and it had decayed fairly soon and disappeared from the face of Earth. If we agree with this idea we have to make two extremely unlikely assumptions. First, illinium which is at the centre of the periodic table has no stable isotopes. Second, the half-lives of its isotopes are much shorter than the age of Earth.

Indeed, illinium neighbours in the periodic system (neodymium and samarium) have many (seven each) natural isotopes with a wide range of mass numbers - from 142 to 154. Any feasible isotope of element 61 would have its mass number in this range. Thus, any illinium isotope proves to be unstable in this range of mass numbers.

The Mattauch law seemed to bury for good the hopes to find element 61 on Earth. But then a gleam of hope appeared. All right, the illinium isotopes are all radioactive. But to what extent? Perhaps the half-lives of some of them are very long. At that time the theory had not learned how to predict half-lives of isotopes. The search for element 61 had to continue in the dark. Physicists believed that only nuclear synthesis could solve the riddle of element 61 the more so as the case of technetium was fresh in their minds.

As if trying to restore the honour of American science, after its setback in 1926 two physicists from the University of Ohio conducted the first experiment on artificial synthesis of element 61 in 1938. They bombarded a neodymium target with fast deutrons (the nuclei of heavy hydrogen). They believed that the resulting nuclear reaction Nd+d 61+n gave rise to an isotope of element 61. Their results were inconclusive but nevertheless they thought that they had obtained an isotope of the new element with the mass number of 144 and the half-life of 12.5 hours.

Again sceptics said that these results were erroneous and not without a reason since nobody could be sure that the neodymium target was ideally pure. The method of identification could hardly be considered reliable, too. Even uncomplicated optical and X-ray spectra evidenced the presence of element 61 as in the study of 1926; the conclusion was made from the radiometric data.

In fact chemistry was not involved in this work and the chemical nature of the mysterious radioactive product was not determined. Therefore, one may ask whether 1938 can be regarded as the actual date of discovery of element 61. It can rather be said that only the consistent efforts to synthesize it started at the time.

As time passed, the range of bombarding particles was extending, targets of other rare-earth elements were used, and the techniques of activity measurements were improved. Reports on other illinium isotopes started to appear in scientific journals. Element 61 was becoming a reality albeit an artificially-created one. Its name was changed to cyclonium in commemoration of the fact that it was produced in a cyclotron but the symbol Cy did not remain for long in box 61 of the periodic table.

Researchers had detected the radioactive "signal" of cyclonium but nobody had seen even a grain of the new element and its spectra had not been recorded. Only indirect evidence of the existence of cyclonium had been obtained.

The history of science of the 20th century knows of many great discoveries and one of the greatest is the discovery of uranium fission under the effect of slow neutrons. The nuclei of uranium-235 isotopes are split into two fragments, each of which is an isotope of one of the elements at the centre of the periodic table. Isotopes of thirty odd elements from zinc to gadolinium can be produced in this way. The yield of the isotopes of element 61 has been calculated to be fairly high-approximately 3 percent of the total amount of the fission products.

But the task of extracting the 3 percent amount proved to be very difficult.

The American chemists J. Marinsky, L. Glendenin and Ch. Coryell applied a new chemical technique of ion-exchange chromatography for separation of the uranium fission fragments.

Special high-molecular compounds known as the ion-exchange resins are employed in this technique for separating elements. The resins act as a sieve sorting up elements in an order of the increasing strength of the bonds between the respective elements and the resin. At the bottom of the sieve the scientists found a real treasure - two isotopes of element 61 with the mass numbers 147 and 149.

At last, element 61 known as illinium, florencium, and cyclonium could be given its final name. According to recollections of the discoverers, the search for a new name was no less difficult than the search for the element itself. The wife of one of them, M. Coryell, resolved the difficulty when she suggested the name promethium for the element. In an ancient Greek myth, Prometheus stole fire from heaven, gave it to man and was consequently put to extreme torture by Zeus. The name is not only a symbol of the dramatic way of obtaining the new element in noticeable amounts owing to the harnessing of nuclear fission by man but also a warning against the impeding danger that mankind will be tortured by the hawk of war, wrote the scientists.

Promethium was obtained in 1945 but the first report was published in 1947. On June 28, 1948, the participants at a symposium of the American Chemical Society in Syracuse had a lucky chance to see the first specimens of promethium compounds (yellow chloride and pink nitrate) each weighing 3mg. These specimens were no less significant than the first pure radium salt prepared by Marie Curie. Promethium was born by the great creative power of science. The amounts of promethium prepared now weigh tens of grams and most of its properties have been studied.

The Mattauch law denied the existence of terrestrial promethium but this denial was not absolute. The search for promethium in terrestrial ores and minerals would be quite in order if promethium had long lived isotopes with half-lives of the order of the age of Earth.

But in this respect nuclear physics proved to be a foe of natural promethium. With each newly synthesized promethium isotope a possible scope for search became increasingly narrow.

The promethium isotopes were found to be short-lived. Among the fifteen promethium isotopes known today, the longest-lived one has a half-life of only 30 years. In other words, when Earth had just formed as a planet not a trace of promethium could exist on it. But what we mean here is the primary promethium formed in the primordial process of origination of elements. What was discussed was the search for the secondary promethium which is still being formed on Earth in various natural nuclear reactions.

Technetium was finally found on Earth among the fragments of spontaneous fission of uranium. These fission products could contain promethium isotopes. According to estimates, the amount of promethium that can be produced owing to spontaneous fission of uranium in the Earth's crust is about 780g, that is, practically, nothing. To look for natural promethium would be tantamount to dissolving a barrel of salt in the lake Baikal and then trying to find individual salt molecules.

But this titanic task was fulfilled in 1968. A group of American scientists including the discoverer of natural technetium P. Kuroda managed to find the natural promethium isotope with a mass number of 147 in a specimen of uranium ore (pitchblende). This was the final step in the fascinating history of the discovery of element 61.

As in the case of technetium, we can name two dates of discovery of promethium.

The first date is the date of its synthesis, 1945. But under the circumstances, synthesis was unconventional (it could be called fission synthesis). The first two promethium isotopes were extracted from the fragments of fission of uranium irradiated with slow neutrons rather than in a direct way as was the case with technetium, which was produced in a direct nuclear reaction. This makes promethium a unique case among all other synthesized elements.

The second date is the date of the discovery of natural promethium, that is, 1968. This achievement is of independent significance as it stretched to the utmost the capabilities of the physical and chemical methods of analysis. Of course, the achievement is of a purely theoretical significance since nobody can hope to extract natural promethium for practical uses.

ASTATINE AND FRANCIUM

ASTATINE (Greek astatos "unstable") – Non-metallic, radioactive element, symbol At, atomic number 85, relative atomic mass 210. It is a member of the halogen group, and is very rare in nature. Astatine is highly unstable, with at least 19 isotopes; the longest lived has a half-life of about eight hours.

FRANCIUM – Radioactive metallic element, symbol Fr, atomic number 87, relative atomic mass 223. It is one of the alkali metals and occurs in nature in small amounts as a decay product of actinium. Its longest-lived isotope has a half-life of only 21 minutes. Francium was discovered and named in 1939 by Marguerite Perey to honour her country.

In July 1925 the British scientist W. Friend went to Palestine but not as a pilgrim. Moreover, he was neither an archaeologist nor a tourist visiting exotic lands. He was just a chemist and his luggage contained mostly ordinary empty bottles which he hoped to fill with samples of water from the Dead Sea, which has the highest concentration of dissolved salts on Earth. Fish cannot live in it and a man can swim in it without any danger of drowning - so high is the density of water in it.

The Sombre Biblical landscapes failed to dampen Friend's hopes for success. His goal was to find in the water of the Dead Sea eka-iodine and eka-caesium which chemists had sought in vain. Sea water contains many dissolved salts of alkali metals and halogens and their concentration in the Dead Sea water must be exceptionally high. The higher the probability that they hide among them the unknown elements, namely the heaviest halogen and the heaviest alkali metal, even if in trace amounts.

Of course, Friend was not entirely original in choosing the direction of his search. As early as the end of the 19th century, a chemist would not hesitate to answer the question where to look for eka-iodine and eka-caesium on Earth. The obvious answer was where natural compounds of alkali metals and halogens are found, that is, in deposits of potassium salts, in sea and ocean water, in various minerals, in deep well water, in some sea algae, and so on. In other words, the field of search was quite wide.

But all the attempts to find eka-iodine and eka-caesium failed and efforts to Friend were no exception.

Now let us turn back to the last decades of the 19th century. When Mendeleev developed the periodic system of elements, it contained many empty slots corresponding to unknown elements between bismuth and uranium. These empty slots were rapidly filled after the discovery of radioactivity. Polonium, radium, radon, actinium, and finally protactinium took their places between uranium and thorium. Only eka-iodine and eka-caesium were late. This fact, however, did not particularly trouble scientists. These unknown elements had to be radioactive since there was not even a hint of doubt that radioactivity was the common feature of elements heavier than bismuth. Therefore, sooner or later radiometric methods would demonstrate the existence of elements 85 and 87.

The natural isotopes of uranium and thorium in long series of successive radioactive transformations give rise to secondary chemical elements. In the first decade of the 20th century, scientists had in their disposal about forty radioactive isotopes of the elements at the end of the periodic system, that is, from bismuth to uranium. These radioelements comprised three radioactive families headed by thorium-232, uranium-235 and uranium-238. Each radioactive element sent its representatives to these families with the only exception of eka-iodine and eka-caesium. None of the three series had links that would correspond to the isotopes of element 85 or 87. This suggested an unexpected idea that eka-iodine and eka-caesium were not radioactive. But why? Nobody dared to answer this question. Under this assumption, it was meaningless to look for these elements in the ores of uranium and thorium which contained all the radioactive elements without exception.

The assumption about stability of eka-iodine and eka-caesium was not confirmed. But all efforts to find isotopes of these elements in the radioactive families met with failure. But there remained one path of investigation which seemed promising. Does a given radioactive isotope have only one or two decay mechanisms? For instance, it emits both alpha and beta particles. If so, the products of decay of this isotope are isotopes of two different elements and the series of radioactive transformations at the place of this isotope experiences branching. This problem was discussed for a long time and for some isotopes this effect seemed to take place.

In 1913, the British scientist A. Cranston worked with the radioelement Ms Th-II (an isotope of actinium-228). This isotope emits beta particles and converts into thorium-228. But Cranston thought that he had detected a very weak alpha decay, too. If that was true the product of decay had to be the long-expected eka-caesium. Indeed, the process is described by

$$\ce{^{228}_{89}Ac} \xrightarrow{\lambda} \ce{^{224}_{87}Fr}$$

But Cranston just reported his observation and did not follow the lead.

Just a year later, three radiochemists from Vienna - S.Meyer, G. Hess, and F. Paneth - studied actinium-227, an isotope belonging to the family of uranium-235. They repeated their experiments and at last their sensitive instruments detected alpha particles of unknown origin. Alpha particles emitted by various isotopes have specific mean paths in air (of the order of a few centimetres). The mean path of the alpha particles in the experiments of the Austrian scientists was 3.5 cm. No known alpha-active isotope had such mean path of alpha particles. The scientists from the Vienna Radium Institute concluded that these particles were the product of alpha decay of the typically beta-active actinium-227. A product of this decay had to be an isotope of element 87.

The discovery had to be confirmed in new experiments. The Austrians were ready for this but soon the World War I started.

They indeed observed alpha radiation of actinium-227 and this meant that atoms of element 87 were produced in their presence. But this fact had to be proved. It was easier to refute their conclusions. Sceptics said that the observed alpha activity was too weak and the results were probably erroneous. Others said that an isotope of the neighbouring element, protactinium, also emitted alpha particles with the mean path close to 3.5 cm. Perhaps, an error was caused by an admixture of protactinium.

Elements 85 and 87 were discovered several times and given such names as dacinum and moldavium, alcalinium and helvetium, or leptinum and anglohelvetium. But all of them were mistakes. The fine sounding names covered emptiness.

The mass numbers of all isotopes in the family of thorium-232 are divided by four. Therefore, the thorium family is sometimes referred to as the 4n family. After division by four of the mass numbers of the isotopes in the two uranium families, we get a remainder of two or three. Respectively, the uranium-238 family is known as the (4n+2) family and the uranium-235 family as the (4n+3) family.

But where is the (4n+1) family? Perhaps, it is precisely in this unknown fourth series of radioactive transformations that the isotopes of eka-iodine and eka-caesium can be found. The idea was not unreasonable but not a single known radioactive isotope could fit into this hypothetical family by its mass number.

Sceptics declared, not without reason, that indeed there had been the fourth radioactive series at the early stages of Earth's existence. But all the isotopes that comprised it including the originator of the series had too short half-lives and hence disappeared from the face of the Earth long ago. The fourth radioactive tree had withered away long before mankind appeared.

In the twenties, theorists attempted to reconstruct this family, to visualize its composition if it had existed. This imaginary structure had positions for the isotopes of elements 85 and 87 (but not for the radon isotopes). But this direction of search did not bring results, too. Perhaps the elusive elements did not exist at all?

But the goal was not that far. But before we start the tale about the realization of the scientists' dreams, let us turn back to the first synthesized element, namely, technetium.

Why was technetium the first? Primarily, because the choice of the target and the bombarding particles was obvious. The target was molybdenum, which could be made sufficiently pure at the time. The bombarding particles were neutrons and deutrons, and accelerators were available for accelerating deutrons. This is why the discovery of technetium manifested the dawn of the age of synthesized elements.

The work on promethium proved more complicated because it belonged to the rare-earth family and the main difficulties were met in determining its chemical nature.

But the task for elements 85 and 87 looked much more formidable. In their attempts to produce eka-iodine the scientists could only have one material for the target, namely, bismuth, element 83. The bombarding particles were a case of Hobson's choice too, only alpha particles could be used. Polonium, which precedes eka-iodine, could not be used as the material for the target. The elements with lower numbers than bismuth could not be used as targets because the scientists lacked appropriate bombarding particles to reach number 85.

Eka-caesium looked totally inaccessible for artificial synthesis. No suitable targets and bombarding particles existed in the thirties. But such is the irony of history that it was precisely element 87 that became the second after technetium reliably-discovered element out of the four missing elements within the old boundaries of the periodic system.

At this point in history, the line of eka-iodine and eka-caesium, which had travelled parallel for such a long time, started to diverge and therefore we shall consider their discoveries separately.

Element 85 was synthesized by D. Corson, C. Mackenzie and E. Segre who worked at Berkley (USA). The Italian physicist Segre by that time had settled in the USA and was the only one in the group who had an experience in artificial synthesis of a new element (technetium). On July 16, 1940, these scientists submitted to the prestigious physical journal Physical Review a large paper entitled "Artificial radioactive element 85". They reported how they had bombarded a bismuth target with alpha particles accelerated in a cyclotron and obtained a radioactive product of the nuclear reaction. The product, most probably, was an isotope of eka-iodine with a half-life of 7.5 hours and a mass number of 211. Segre and his co-workers performed fine chemical experiments with the new element produced in negligible amounts and found that it was similar to iodine and exhibited weakly metallic properties.

The results seemed convincing enough. But the new element remained nameless for the time being. Further work on eka-iodine had to be delayed as the war started. It was resumed only in 1947 and the same group announced synthesis of another isotope with a mass number of 210. Its half-life was somewhat longer but still only 8.3 hours. Later it was found to be the longest-lived isotope of element 85. It was produced with a similar technique as the first isotope though the energy of the bombarding alpha particles was somewhat higher. As a result the intermediate composite nucleus (^{209}Bi+) emitted three rather than two neutrons and, hence, the mass number of the isotope was lower by 1. Only now the new element was given the name "astatine" from the Greek for "unstable" (the symbol At).

But in the interval between the syntheses of the isotopes ^{211}At and ^{210}At a remarkable event occurred. The scientists from the Vienna Radium Institute B. Karlik and T. Bernert managed to find natural astatine. This was an extremely skilful study straining to the utmost the capacity of radiometry. The work was crowned with success and element 85 was born for the second time. As in the cases of technetium and promethium, we can name two dates in the history of astatine, namely, the year of its synthesis (1940) and the year of its discovery in nature (1943).

But when Segre and his coworkers were preparing for irradiating a bismuth target with alpha particles the scientific community had known about the discovery of eka-caesium for more than a year. Transactions of the Paris Academy of Sciences published a paper headed "Element 87: Ack formed form actinium" and dated January 9, 1939. Its author was M. Perey, the assistant of the eminent radiochemist Debierne who had announced his discovery of actinium forty years earlier.

Marguerite Perey did not invent any fundamentally new methods and did not indulge in any vague and complicated speculations about possible sources of natural eka-caesium. In 1938 she came upon a paper published in 1914. The paper was signed by the Austrian chemists Meyer, Hess and Paneth. Perey attempted to prove their ideas. She obtained a carefully purified specimen of actinium-227. This isotope has a high beta-activity but sometimes it emits alpha particles, too. The mean path of such particles in air is 3.5 cm. This alpha radiation is by no means due to protactinium as the actinium specimen was sufficiently purified. Since alpha particles are emitted, the eka-caesium isotope with a mass number of 223 must continuously be accumulated in the specimen. A series of experiments definitely demonstrated that, indeed, some substance with a half-life of 21 min is accumulated in the actinium specimen. Now it is the turn of chemical analysis to prove that this substance is a new element. Its properties proved to be similar to those of caesium. Perey named the new element francium in honour of her country. Only for a short period it was called actinium K (AcK) in accordance with the old nomenclature of radioelements.

The first description given by Perey to the new born element was extremely brief: the element is formed with alpha decay of actinium-227 in the reaction

$$\ce{^{227}_{89}Ac} \xrightarrow{\lambda} \ce{^{221}_{85}Fr}$$

and it is alpha-active with a half-life of 21 min. Then she spent several months studying its chemical properties and demonstrated convincingly that francium is similar to caesium in all its characteristics.

None of the natural radioactive elements had such a short half-life, even the artificially synthesized element 85 had a half-life measured in hours. There were hopes to find other natural isotopes of francium with longer half-lives. But in fact francium-223 proved to be the only francium isotope found on Earth.

The only remaining path to success was synthesis but it proved very difficult. More than ten years passed after the discovery of Perey when francium isotopes were, artificially synthesized. The nuclear reaction giving rise to the francium isotope with a mass number of 212 can be written in short as

$$_{92}^{232}U(p, 6p\ 21_n)_{87}^{221}Fr$$

This reaction is the fission of uranium nucleus by protons accelerated to very high energies. When such a fast proton hits uranium nucleus, it produces something like an explosion with ejection of a multitude of particles, namely, six protons and 21 neutrons. Of course, the reaction is not due to a blind chance but is based on careful theoretical predictions. Uranium may be replaced with thorium. The reaction product, francium-212, for some time was considered to be the longest-lived isotope (a half-life of 23 minutes) but later the half-life was found to be only 19 minutes.

Artificial synthesis of francium is much more difficult and less reliable method than extraction of francium as a product of decay of natural actinium. But natural actinium is rare. What to do? A current method is to irradiate the main isotope of radium with a mass number of 226 (its half-life is 1622 years) with fast neutrons. Radium-226 absorbs a neutron and converts into radium-227 with a half-life of about 40 minutes. Its decay gives rise to pure actinium-227 whose alpha decay in its turn produces francium-223.

The symbols At and Fr were permanently installed in boxes 85 and 87 of the periodic table and their properties proved to be exactly the same as predicted from the table. But in comparison with their unstable mates born by nuclear physics, technetium and promethium, their position is clearly unfavourable.

According to estimates, the 20km thickness of the Earth crust contains approximately 520g of francium and 30g of astatine (this is an overestimation in some respects). These amounts are of the same order as the terrestrial "resources" (quotation marks are more than suitable here) of technetium and promethium. Are we probably making a mistake when we talk condescendingly about astatine and francium? Not at all. Technetium and promethium are produced in large amounts, kilograms and kilograms of them. The fact is that technetium and promethium have much longer half-lives and can therefore be accumulated in large amounts. But accumulation of astatine and francium is just unfeasible. In fact, each time their properties have to be studied they have to be produced anew.

In the radioactive families, the isotopes of astantine and francium are placed not on the principal pathways of radioactive transformations but at the side branches. Here is the branch on which natural francium is born:

$$^{227}_{89}AC \Big\langle \begin{matrix} \xrightarrow{\beta} {}^{227}_{90}AC \\ \xrightarrow{\alpha} {}^{223}_{87}Fr \end{matrix}$$

The isotope $^{227}_{89}AC$ in 99 cases out of 100 emits beta particles and only in one case it undergoes alpha decay.

The situation is even less easy in the case of the branches responsible for the formation of astatine:

$$^{218}_{84}Po \Big\langle \begin{matrix} \xrightarrow{\alpha} {}^{214}_{82}Pb \\ \xrightarrow{\beta} {}^{218}_{85}At \end{matrix} \quad ^{216}_{84}Po \Big\langle \begin{matrix} \xrightarrow{\alpha} {}^{212}_{84}Pb \\ \xrightarrow{\beta} {}^{216}_{85}At \end{matrix} \quad ^{215}_{84}Po \Big\langle \begin{matrix} \xrightarrow{\alpha} {}^{211}_{82}Pb \\ \xrightarrow{\beta} {}^{215}_{85}At \end{matrix}$$

What may be said about these branches? The producers of natural astatine (the polonium isotopes) are by themselves extremely rare. For them alpha decay is not just predominant but practically the only radioactivity mechanism. Beta decays for them seem something like a mishap as can be clearly seen from the following data.

There is only one beta decay event per 5000 alpha decays of polonium-218. Things are even sadder for polonium-216 (1 per 7000) and polonium-215 (1 per 200000). The situation speaks for itself. The amount of natural francium on Earth is larger. It is produced by the longest-lived actinium isotope ^{227}Ac (a half-life of 21 years) and its content is, of course, much higher than that of the extremely rare polonium isotopes capable of producing astatine.

13

Transuranium elements

Transuranium elements are all the elements whose numbers are higher than 92, that is, the elements that directly follow uranium. Now 26 such elements are known. How many more transuranium elements can be found? The answer is still unknown. This is one of the fascinating mysteries of science.

Though the first transuranium element, neptunium, (No. 93) was born not so long ago, in 1940, the question about possible existence of such elements was raised much earlier. Mendeleev did not ignore it either. He believed that even if the transuranium elements would be found on Earth, their number will be limited. This was his opinion in 1870. For more than 25 years, the problem remained open. Every year saw several erroneous reports on discoveries of new elements but not once the element in question had an atomic mass greater than that of uranium. It seemed axiomatic that uranium was the last element in the periodic system though nobody could say why.

But when radioactivity was discovered thorium and uranium, the heaviest elements in the Mendeleev table, were found to possess this property. It would logically seem that the transuranium elements had existed in nature in past but, being highly unstable, had decayed to other, known elements.

This simple explanation had a hidden trap, namely, the possible half-lives of even the nearest right hand neighbours of uranium were quite unknown. Nobody could state with certainty that these hypothetical elements were less stable than uranium and thorium. Thus, it would be reasonable to look for natural transuranium elements.

Years passed and occasionally allegedly successful discoveries of the first transuranium element were reported in scientific journals. As theoretical physics developed, it repeatedly attempted to explain the break off of the periodic system at uranium. Many of these explanations were fascinating but none convincing. In other words, in the twenties of this century the question of transuranium elements looked as vague as in the last quarter of the 19th Century.

One amazing hypothesis appeared, however, against this dismal background though at first scientists treated it with suspicion. Only 40 years later, the hypothesis found a new meaning. It was put forward in 1925 by the German scientist R. Swinne who looked for the transuranium elements in a peculiar material - a dust of space origin collected on the icefields of Greenland. A sample of the dark powder was given to the Stockholm museum by the well-known polar explorer E. Nordenskjold in the eighties of the last century. Swinne hoped to find in this powder traces of transuranium elements with the numbers 106-110 and in one of his reports, he even mentioned that he had recorded an X-ray spectrum containing lines that, in his opinion, corresponded to element 108. But nobody believed him and he himself discontinued his work.

Swinne made a theoretical study of the variation of various properties of radioactive elements and, in particular, half-lives. He came to the conclusion that the elements directly following uranium had to have short half-lives. But the elements with the numbers in the ranges between 98 and 102 and between 108 and 110 could be expected to have sufficiently long half-lives.

Where to look for them? Swinne suggested that the best bet would be not terrestrial minerals but space objects. This is why he studied the dust of space origin collected in Greenland. All this was quite fascinating but not substantiated and therefore looked like doomed for oblivion.

Now we come to the point in time when the words "transuranium elements" started to be linked with the word "synthesis".

Paradoxical as it seems, the attempts to synthesize new elements (namely, transuranium elements) had started a few years before technetium was produced. The stimulus for this work was the discovery of neutron. The scientists regarded this chargeless elementary particle as possessing infinite penetrating capacity and being capable of producing a wide variety of transformations of all kinds of elements. Thus, all laboratories that had neutron sources started to bombard with neutron targets made of various materials including uranium. Especially active in this work was the Italian physicist E. Fermi who was the leader of a group of young enthusiasts at the University of Rome.

They detected some new activity in irradiated uranium. As they irradiated uranium-238, it absorbed neutrons converting into an unknown uranium isotope with a mass number of 239. Since the isotope had an excess of neutrons, it exhibited a definite tendency to beta decay. If the left hand side of the reaction equation is $^{239}U-\beta$ than the right hand side is necessarily $^{239}93$.

Fermi and his young coworkers argued in approximately this way (though not very clearly as many concepts of nuclear physics at the time were not sufficiently developed). Now chemical verification was needed to prove the synthesis of the first transuranium element. It had to be demonstrated that the activity induced by neutrons in uranium did not belong to any of the preceding elements. This was established within the limits of the capacity of radiochemistry. Thus, Fermi and his group had in their hands a new element, a transuranium one and one that was the first to be discovered owing to the nuclear synthesis (all this happened in 1934). Fermi and his group, however, were not completely sure of their results. Meanwhile the news about the new element leaked to the press and the discovery was embellished with non-existent details such as Fermi presenting the queen of Italy with a test-tube containing a dissolved salt of element 93. A lot of such false sensations were published in press while the group continued to assess the results obtained through irradiation of uranium with neutrons.

They extracted several beta-active substances from the uranium target. Two of them were chemically peculiar as they could be precipitated with manganese (IV) oxide easier than the elements preceding uranium. This observation suggested that element 93 was eka-rhenium - a manganese analogue. It was named auzonium (Ao). Being beta active it could convert into the next element with Z=94 known as hesperium (Hs). Fermi described this series of nuclear transformations in the following way.

$$^{238}_{92}U + n \rightarrow ^{239}_{92}U \xrightarrow{\beta^-} ^{239}_{93}Ao \xrightarrow{\beta^-} ^{239}_{94}HS$$

This series was continued further by the German scientists O. Hahn, L. Meitner and F. Strassmann, highly experienced radiochemists, particularly O. Hahn who had made his name having had discovered several radioelements. As a result of careful studies, the number of new transuranium elements increased by three (including element 97):

$$_{94}HS \xrightarrow{\beta^-} _{95}EKaIr \xrightarrow{\beta^-} _{96}EKapt \xrightarrow{\beta^-} _{97}EKaAu$$

The prefix Eka means that the respective transuranium elements were considered to be analogues of iridum, platinum and gold from the sixth period of the periodic system. But it was precisely here that a serious mistake was made, which took quite a time to be found. The properties of the nearest transuranium elements were, in fact, quite different.

The history of science knows of many marvellous insights which seemed at first quite unsubstantiated. One of them was the idea put forward by I. Noddack back in 1934 that when uranium was bombarded with neutrons the uranium nuclei did not convert into new elements at all, rather, they were split into fragments which were the nuclei of lighter, known elements. Her colleagues made light of Noddack's idea and Hahn's comments were especially ironic. But his irony turned to be the irony of fate.

Meanwhile, other scientists tried to ascertain what happened to uranium under neutron bombardment. I. Joliot Curie and her coworker, the Serbian physicist P. Savich, particularly carefully analysed an irradiated uranium target. Among the resulting activities they detected traces of a chemical element whose properties were very similar to those of actinium, that is, an element preceding uranium, rather than following it in the periodic table. Soon it was found to have more in common with lanthanum than with actinium. Thus, one of the products obtained after bombardment of uranium with slow neutrons was similar to lanthanum.

If I. Joliot-Curie and Savich had not drawn a line at cautiously stating that the unknown element was similar to lanthanum but had definitely proven that it was lanthanum they would have become the authors (or, at least, coauthors) of one of the greatest discoveries of the 20th century. (It would be in order here to recall that lanthanum has the number 57 and uranium the number 92 and to recall the idea of I. Noddack, too.) This seemed more than improbable. But facts remained facts. The results of I. Joliot-Curie and Savich looked so convincing that O. Hahn took it upon himself to verify them, the very same O.

Hahn who was an ardent opponent of these results. This meant that he had started to question his former opinions.

Hahn, together with his coworker Strassmann, reproduced the experiments of the French scientists whom he so recently had regarded as his opponents. Almost all the results were confirmed. The uranium target contained isotopes of lanthanum and its preceding neighbour in the periodic system, barium. As a chemist, Hahn could not doubt this. As a physicist, he was baffled by the fact.

The fact was that under neutron bombardment, uranium nuclei seemed to split into two fragments and these fragments were the nuclei of the isotopes of elements belonging to the centre of the periodic system. Nuclear physics had never encountered such a phenomenon. But facts had to be faced and the German scientists concluded that uranium nuclei were capable of breaking down under neutron bombardment.

This happened on December 23, 1938. The scientists immediately reported their discovery. Later Hahn reminisced that after posting the report it all had seemed so improbable to him that he had wished that he could take the letter back from the post box.

The improbable proved to be right. A few days later a letter from Hahn was received by L. Meitner who had worked with him for many years. She, together with her nephew, the physicist O. Frisch, attempted theoretical treatment of this phenomenon.

To a certain extent, nuclei can be likened to drops of liquid and scientists have repeatedly tried to draw an analogy between the properties of a nucleus and those of a drop of liquid. If we transfer a sufficient energy to a drop and make it move it can break down to smaller drops. If a nucleus is excited (by a neutron, say) then it can also split into smaller fragments. Gradually, a uranium nucleus is deformed, it elongates, narrowing appears in it, and, finally, it splits into two parts. This is how Meitner and Frisch described the process of splitting of the uranium nuclei. They wrote that the process was remarkably similar to division of bacterial cells by which they propagate and suggested naming the effect "nuclei fission".

A uranium nucleus splits into two fragments liberating an enormous amount of energy in the process. Other products of fission were free neutrons. They could hit other uranium nuclei leading to their fission and so on. Under favourable conditions a chain fission reaction could occur in a uranium lump producing a nuclear explosion of immense power. As early as 1940, the Soviet scientists Ya. Zel dovich and Yu. Khariton developed a rigorous theory of the chain fission reaction. Man mastered a process which, apparently, was unknown in nature. This was the most comprehensive process of transformation of elements man had ever encountered. The fragments of uranium fission were found to contain isotopes of 34 elements, from zinc (number 30) to gadolinium (number 64). The fission reaction proved to be a veritable factory of radioactive isotopes.

Uranium fission caused by neutrons was forced or artificial. Not each uranium nucleus could be split and not each neutron could produce fission. When scientists had studied the fission mechanism in more detail, they understood that the intensity of fission was higher under the effect of slow neutrons and if the uranium isotope with a mass number of 235 was used, the other uranium isotope, uranium-238, experienced fission only when bombarded by fast neutrons. Can there be a natural process similar to artificial uranium fission? N. Bohar thought about that and put forward a hypothesis about possible spontaneous uranium fission (without external energy being transferred to the nuclei).

The Soviet scientists G. Flerov and K. Petrzhak attempted an experimental verification of this hypothesis. But how to establish that fission of the uranium nuclei was really spontaneous? Random neutrons of cosmic rays getting into the laboratory could distort the results of experiments. This is why one autumn midnight of 1940 Flerov and Petrzhak went down to one of the deepest stations of Moscow underground railway. There, tens of metres under the surface of earth, the harmful effect of cosmic rays could be escaped. The same night they obtained the final proof of the existence of a new type of radioactive transformations, namely, spontaneous fission of nuclei (they worked only with uranium-238). Later many isotopes of heavy elements (thorium and, particularly, transuranium elements) were found to exhibit this mechanism of radioactive decay. At present science knows of about a hundred nuclei of various elements capable of spontaneous fission. The mechanism of spontaneous fission is similar to that of fission under neutron bombardment.

We now know enough to embark of the tale of the discoveries of individual transuranium elements since it is just in this range of elements that spontaneous fission plays a very significant part.

The history of transuranium elements covers forty years and during this, by modern standards, fairly long period, scientists managed to take fifteen steps beyond uranium up to element 107. If we take a frame of reference and plot the numbers of elements from 1 to 92 along the horizontal axis and the years of their discovery along the vertical axis the resulting plot will look like a seismogram of a catastrophic earthquake. A similar plot for the transuranium elements is a comparatively smoothly rising line exhibiting distinct peaks. Each new synthesis of a transuranium element meant an increase in the atomic number by one (with a single exception).

The history of syntheses saw its periods of breakthroughs and slack periods. The first breakthrough period was from 1940 to 1945 when four transuranium elements were synthesized, namely, neptunium (Z=93), plutonium (z=94), americium (z=95) and curium (z=96). The period till 1949 was a slack time and no new elements were discovered. In the next breakthrough period from 1949 to 1952 four more transuranium elements were added to the periodic system, namely berklium (z=97), californium (Z=98), einsteinium (z=99) and fermium (z=100). In 1955, fifteen years after the synthesis of the first transuranium element, one more element, mendelevium (z=101), was synthesized. The next 25 years saw much less syntheses and only six new elements appeared in the periodic system. Here scientists encountered an entirely new situation and many former criteria for evaluating discoveries of elements proved inapplicable.

This changing pattern is by no means random, all the breakthroughs and failures had their quite objective causes. They will be apparent when we discuss syntheses of transuranium elements one by one starting from the first one, neptunium.

NEPTUNIUM

NEPTUNIUM – Silvery, radioactive metallic element of the actinide series, symbol Np, atomic number 93, relative atomic mass 237.048. It occurs in nature in minute amounts in pitchblende and other uranium ores, where it is produced from the decay of neutron-bombarded uranium in these ores. The longest-lived isotope, Np-237, has a half-life of 2.2 million years. The element can be produced by bombardment of U-238 with neutrons and is chemically highly reactive.

It was first synthesized in 1940 by US physicists E McMillan and P Abelson, who named it for the planet Neptune comes after Uranus. Neptunium was the first transuranic element to be synthesized.

Of course, Fermi never presented the queen of Italy with a test tube containing a salt of the first transuranium element. It is no more than a typical newspapermen's copy. But it is true that Fermi had in his hands element 93 though it could not be proved at the time. In his experiments the uranium target consisted of two isotopes, namely, uranium-238 and uranium-235. The latter underwent fission under the effect of slow neutrons giving rise to fragments which were the nuclei of the elements belonging to the central part of the periodic system. They greatly complicated the chemical situation but this was understood only when fission was discovered.

But uranium-238 absorbed neutrons converting into uranium-239, a new isotope of uranium. This beta-active isotope gave rise to an isotope of the first transuranium element with an atomic number of 93. This was just what Fermi and his group thought. But the future neptunium was hard to distinguish among the multitude of fragments. This is why the experiments in mid-thirties yielded no results.

The discovery of Hahn and Strassmann decisively stimulated actual synthesis of transuranium elements. To start, a reliable technique was needed for detection of the atoms of element 93 in a mass of fission fragments. As the masses of these fragments were comparatively small, they had to travel longer distances (had longer paths) than the atoms of element 93 with a large mass.

Thus went the argument of E. McMillan, an American physicist from the University of California. Back in the spring of 1939, he started to analyse the distribution of uranium fission fragments along their paths. He managed to obtain a sample of fragments whose path was very short and in this sample he found traces of a radioactive substance with a half-life of 2.3 days and a high radiation intensity. Other parts of the fission fragments did not exhibit such activity. McMillan demonstrated that this unknown substance was a fission product of a uranium isotope which was also found in the short-path fragments. Thus, the reaction sequence first suggested by Fermi was written as

$$^{238}_{92}U + n \rightarrow {}^{239}_{92}U \xrightarrow{\beta^-} 239$$

Now the search was no longer conducted in darkness. Chemical analysis had then to be the final step in verification of the new element. On summer vacations McMillan invited his friend, the chemist P. Abelson, and this visit played a crucial part in the discovery of element 93. Together they established the chemical nature of the new element with a half-life of 2.3 days. The element could be chemically separated from thorium and uranium, though in some aspects it was similar to them. But the new element was in no way similar to rhenium. This finally refuted the hypothesis that element 93 had to be eka-rhenium.

At the beginning of 1940, the Physical Review Journal reported the real discovery of element 93. It was named neptunium after the planet that is beyond Uranus in the solar system (there is some analogy to the periodic system where neptunium follows uranium).

Synthesis of neptunium exhibited a significant feature which was to prove typical for syntheses of all transuranium elements (and other synthesized elements, too). First, one isotope with a certain mass number was synthesized. For neptunium, this was neptunium-239. From that time it became a rule to date a discovery of a new transuranium element by the time of reliable synthesis of its first isotope. But sometimes this isotope proved to be so short-lived that it was difficult to subject it to physical and chemical analyses let alone find a useful application for it. A study of a new element would best be conducted with its longest-lived isotope. In the case of neptunium, this was neptunium-237 synthesized in 1942 in the following reaction:

$$^{238}_{92}U(n,2n)^{237}_{92}U \xrightarrow{\beta^-} ^{237}_{93}NP$$

This isotope has a half-life of 2.2×10^6 years. However, its synthesis involves great technical difficulties. Therefore, all the initial studies of the properties of neptunium were performed with its third isotope, neptunium-238, synthesized in the nuclear reaction Therefore, the history of transuranium elements notes also the date of synthesis of the isotope that is most convenient for analysis but which is by no means always the longest-lived one.

Starting from neptunium, the American scientists for a long time played a leading part in discoveries of transuranium elements. This can easily be explained by the fact that USA hardly experienced the hardships of the World War II. It should be noted, however, that in 1942 element 93 was independently synthesized by the German physicist K. Starke.

In 1944, a weighable amount (a few micrograms) of neptunium was synthesized. Now it is produced in tens of kilograms in nuclear reactors.

Thirteen neptunium isotopes are currently known. One of them, neptunium-237 was found in 1957 in nature. This is another example when a previously synthesized element was found in nature and for which two discovery dates can be given (as for technetium, promethium, astatine and francium).

PLUTONIUM

PLUTONIUM – Silvery-white, radioactive, metallic element of the actinide series, symbol Pu, atomic number 94, relative atomic mass 239.13. It occurs in nature in minute quantities in pitchblende and other ores, but is produced in quantity only synthetically. It has six allotropic forms and is one of the three fissile elements (elements capable of splitting into other elements - the others are thorium and uranium). The element has awkward physical properties and is the most toxic substance known.

Because Pu-239 is so easily synthesized from abundant uranium, it has been produced in large quantities by the weapons industry.

It has a long half-life (24,000 years) during which time it remains highly toxic.

Plutonium is dangerous to handle, difficult to store and impossible to dispose of. It was first synthesized in 1940 by Glenn Seaborg and his team at the University of California at Berkeley, by bombarding uranium with deuterons; this was the second transuranic element to be synthesized, the first being neptunium.

The isotope neptunium-239 was beta-active and had to convert regularly into an isotope of the next element (No. 94). McMillan and Abelson, of course, hoped to discover this element, too, but their dream did not come true. As found later, the isotope of element 94 with a mass number of 239 has a long half-life and its activity is low. The discoverers of neptunium only detected alpha particles of an unknown origin (later found to be emitted precisely by element 94) and discontinued their work.

The work on the synthesis of element 94 was headed by the famous American scientist G. Seaborg whose group discovered many transuranium elements. During the winter of 1940-1941, they studied the nuclear reaction which gave rise to the isotope neptunium-238. An alpha-active substance accumulated with time in the reaction product. The scientists extracted this substance and found that it was an isotope of element 94 with a mass number of 238 and a half-life of 50 years. The new element was named plutonium after the respective planet of the Solar system.

But once more this isotope was not the longest-lived one. The longest-lived isotope with a mass number of 244 and a half-life of 8.3×10^7 years was found only in 1952. The decisive progress in the study of plutonium was due to the isotope plutonium-239 synthesized in spring of 1941. First, it was long-lived (a half-life of 24,360 years) and second, the intensity of its fission under the effect of slow neutrons was, much higher than that of uranium-235. This was the decisive factor for its use in nuclear weapons. Therefore, an especially careful study was made of the physical and chemical properties of this element. As a result, plutonium became one of the best-studied elements of the periodic table. Moreover, plutonium-239 could be used as a target for syntheses of next transuranium elements. All this became widely known only at the end of the forties when much of the work on nuclear energy was declassified. This was an unusual feature for the history of elements that discoveries of new elements were kept secret for some time.

The efforts invested into the work on plutonium were so intense that as early as August 1942 weighable amounts of it were prepared (the fastest work in the history of synthesized elements). In our days plutonium is produced in quantities that are much greater than those of many stable elements found on Earth. A total of 17 isotopes of plutonium are currently known.

As in the case of neptunium, the plutonium-239 isotope was found in uranium minerals, of course, in symbolic amounts. It is produced in uranium under the effect of natural neutrons. Thus, plutonium serves as a kind of the natural upper boundary of the periodic system and we can speak about two dates of its discovery.

AMERICIUM AND CURIUM

AMERICIUM – Radioactive, metallic element of the actinide series, symbol Am, atomic number 95, relative atomic mass 243.13; it was first synthesized in 1944. It occurs in nature in minute quantities in pitchblende and other uranium ores, where it is produced from the decay of neutron-bombarded plutonium, and is the element with the highest atomic number that occurs in nature. It is synthesized in quantity only in nuclear reactors by the bombardment of plutonium with neutrons. Its longest-lived isotope is Am-243, with a half-life of 7,650 years.

The element was named by Glenn Seaborg, one of the teams who first synthesized it in 1944, after the United States of America. Ten isotopes are known.

CURIUM – Synthesized, radioactive, metallic element of the actinide series, symbol Cm, atomic number 96, relative atomic mass 247. It is produced by bombarding plutonium or americium with neutrons. Its longest-lived isotope has a half-life of 1.7×107 years.

It is, perhaps, the only occasion in the history of transuranium elements that an element with a higher number (z=96) was identified earlier than its predecessor (z=95). In July 1944, the cyclotron of the University of California, which had already revealed to the world several synthesized elements, including plutonium, was geared to synthesize new transuranium elements. Seaborg and his coworkers bombarded a plutonium-239 target with accelerated alpha particles. One can readily reckon that as the alpha particle (the helium nucleus) has a charge of two, the reaction product could be an isotope of element 96, provided that neutrons were emitted from the resulting nuclei. If the process mechanism was such that protons were emitted, rather than neutrons, then an isotope of element 95 could be synthesized. Indeed, various radioactive substances were produced in the plutonium target and it was difficult at first to identify "who was who". Only skilful chemical analysis revealed that the mixture definitely contained the isotope $^{242}96$. To verify the discovery the same isotope, plutonium-239, was bombarded with a high-intensity neutron beam so that the following chain of reactions took place:

$$^{239}Pu + n \rightarrow {}^{240}Pu + n \rightarrow {}^{241}Pu \xrightarrow{\beta^-} {}^{241}95 + n \rightarrow {}^{242}95 \xrightarrow{\beta^-} {}^{242}96$$

After absorption of neutrons, plutonium converted into element 95 via beta decay and this element obsorbed a neutron and converted into element 96.

This final product was similar to that which the scientists had assumed to be the isotope of element 96 with a mass number of 242. The newly discovered element was named curium after the Curies. Another factor prompted this name. In the Mendeleev table, element 96 was regarded as an analogue of gadolinium belonging to the rare-earth series the history of which had been started by J. Gadolin. In their turn, the Curies were the pioneers of the study of radioactivity whose development produced such amazing results.

In January 1945, elements 95 was extracted from plutonium bombarded with neutrons. The element was named americium in honour of America (and owing to its similarity to europium from the rare-earth series).

Though the researchers had accumulated considerable experience in syntheses, the difficulties involved in producing americium and curium proved unusually great. It took a long time to distinguish definitely between americium-241 and curium-242. Both isotopes proved to be not the longest-lived ones. The longest-lived isotopes were americium-243 (a half-life of 7950 years) and curium-247 (a half-life of 1.64×10^7 years), which were only synthesized in the fifties. The total of 11 americium isotopes and 13 curium isotopes are currently known. Here are a few more events in the history of these elements. Pure americium was extracted in 1945 and in 1951 it was prepared in a metallic form. The same year metallic curium was prepared.

The discovery of curium ends the first breakthrough period in the history of transuranium elements. The discoveries of neptunium, plutonium, americium, and curium were of great significance for science. It was for the first time that scientists artificially extended the boundaries of the periodic system. The properties of these elements proved to be quite different from those expected and chemists had to start seriously thinking how best to fit them into the periodic system.

BERKELIUM

BERKELIUM – Synthesized, radioactive, metallic element of the actinide series, symbol Bk, atomic number 97, relative atomic mass 247.

It was first produced in 1949 by Glenn Seaborg and his team, at the University of California at Berkeley, USA, after which it is named.

Synthesis of americium and curium was promoted by ready availability of plutonium-239. Scientists soon learned how to produce it in large quantities and therefore manufacture of plutonium targets was no problem. Further progress depended on the ability to synthesize americium and curium in sufficient amounts. This took several years. But it was not the only obstacle on the way to new transuranium elements. When a nuclear reaction is written as an equation on paper, it looks amazingly simple but only experts can appreciate the enormous difficulties involved. Researchers had to work out the tiniest details of experiments and to find the optimum conditions for nuclear reactions. They had to perform careful theoretical calculations to predict the types of radioactive transformations of synthesized isotopes and their probable half-lives. Unfortunately, nuclear physicists had no such great assistance as that given to chemists by the wonderful classification of the periodic system. The lull in the discoveries of transuranium elements continued for five years. One more factor should be noted in this respect. Americium and curium have such high activities that it would be deadly dangerous to work with them in the open. Special equipment was needed for such, so-called hot laboratories.

$$^{241}Am(\lambda, 2n)^{243}97$$

At the end of 1949, the group of Seaborg managed to prepare an americium target and to bombard it with alpha particles.

285

The resulting nuclear reaction, as predicted by theorists, was ^{241}Am (α, 2n) ^{243}97. The new element was named berkelium (BK) in honour of Berkley (California) and in connection with the chemical analogy of element 97 to the rare-earth element terbium (recall the village of Ytterbu that gave names to several rare-earth elements). Among the nine currently known berkelium isotopes, the longest-lived one is berkelium-247 (a half-life of 1380 years) which was synthesized in 1956. Two years later a weighable quantity of berkelium was accumulated and in 1971 metallic berkelium was obtained. The difficulties involved in preparation of berkelium are dramatically illustrated by the fact that 8g of plutonium-239 that had been bombarded with neutrons for five years in a nuclear reactor yielded just a few micrograms of berkelium. The further researchers went into the region of transuranium elements, the smaller the quantities of new elements they had to work with.

CALIFORNIUM

CALIFORNIUM – Synthesized, radioactive, metallic element of the actinide series, symbol Cf, atomic number 98, relative atomic mass 251. It is produced in very small quantities and used in nuclear reactors as a neutron source. The longest-lived isotope, Cf-251, has a half-life of 800 years.

It is named after the state of California, where it was first synthesized in 1950 by Glenn Seaborg and his team at the University of California at Berkeley.

Seaborg and his coworkers synthesized element 98 very soon after berkelium. In January-February 1950 they carried out the calculated nuclear reaction ^{242}Cm(λ, n)^{245}98 and named the new element californium in honour of the state of California and the University of California. Moreover, element 98 was an analogue of the rare-earth element dysprosium (difficult to reach) and in the last century to reach California was as difficult as to extract dysprosium from a mixture of rare earths. Fourteen californium isotopes are currently known. The longest-lived one is californium-251 synthesized in 1954 (a half-life of 900 years). Californium was obtained in weighable quantities in 1958 and metallic californium was produced in 1971.

EINSTEINIUM AND FERMIUM

EINSTEINIUM – Synthesized, radioactive, metallic element of the actinide series, symbol Es, atomic number 99, relative atomic mass 254.

It was produced by the first thermo-nuclear explosion, in 1952, and discovered in fallout debris in the form of the isotope Es-254 (half-life 20 days). Its longest-lived isotope, Es-254, with a half-life of 276 days, allowed the element to be studied at length. It is now synthesized by bombarding lower-numbered transuranic elements in particle accelerators. It was first identified by A. Ghiorso and his team who named it in 1955 after Albert Einstein, in honour of his theoretical studies of mass and energy.

FERMIUM – Synthesized, radioactive, metallic element of the actinide series, symbol Fm, atomic number 100, relative atomic mass 257. Ten isotopes are known, the longest-lived of which, Fm-257, has a half-life of 80 days. Fermium has been produced only in minute quantities in particle accelerators.

It was discovered in 1952 in the debris of the first thermonuclear explosion. The element was named in 1955 in honour of Italian-US physicist Enrico Fermi.

After synthesis of californium, scientists in America (and in other countries) started a serious reassessment of their plans. They asked whether it was reasonable to plan for syntheses of heavier transuranium elements in the foreseeable future.

Indeed, there were no practicable methods for accumulation of berkelium and californium in sufficient quantities to prepare targets to be bombarded by alpha particles as a means of synthesizing elements 99 and 100. This was due to short half-lives of berkelium and californium measured in hours and minutes (long-lived isotopes were unknown at the time). There was only one more or less feasible method, namely, to bombard plutonium with a high-intensity neutron beam but then the results would be obtained only many years later.

Of course, it would be desirable to obtain such a high-intensity neutron beam that would solve all the problems at once. If uranium or plutonium could capture a large number of neutrons in a short period, they would convert into very heavy isotopes, for instance,

Or	$^{238}_{92}U + 15n \rightarrow {}^{253}_{92}U$
	$^{238}_{92}U + 17n \rightarrow {}^{255}_{92}U$

It had long been known that nuclei get rid of excess neutrons by converting them into protons, by way of beta decay. These chains of successive beta transformations can prove to be so long that they will lead to the formation of isotopes of elements 99 and 100.

But according to calculations, the intensities of neutron fluxes in nuclear reactors were too low to sustain such reactions. Moreover, theorists predicted short half-lives for the isotopes of elements 99 and 100.

On November 1, 1952, the USA exploded a thermonuclear bomb over the atoll Eniwetok in the Pacific. A few hundred kilograms of the soil from the explosion site was collected with all possible precautions and taken to the USA. A group of scientists headed by Seaborg and Giorso carefully studied this radioactive debris. It was found to contain a variety of radioactive isotopes of transuranium elements including two isotopes which could be nothing else but isotopes of elements 99 or 100.

The intensity of neutron fluxes during the thermonuclear explosion proved to be much higher than it had been expected. This made possible the processes of neutron capture by uranium discussed above. Uranium-253 and Uranium-255 emitted 7 and 8 beta particles, respectively, and converted into isotopes $^{253}99$ and $^{255}100$ of elements 99 and 100. Their half-lives proved to be short but sufficient for analysis (20 days and 22 hours.)

New elements were named einsteinium (after A. Einstein) and fermium (after E- Fermi) Their long-lived isotopes ^{254} Es (a half-life of 270 days) and ^{252} Fm (a half-life of 80 days) were synthesized much later under laboratory conditions.

Thus, the discoveries of einsteinium and fermium were, so to say, unplanned.

The eternal question "What next?" now seemed even more difficult to answer. It was quite clear that the greater the atomic number Z, the shorter the isotope half-live. It was thought that for the elements with Z>100 half-lives would be measured in seconds. It was unthinkable to accumulate these isotopes in quantities sufficient for analysis. Until that time new transuranium elements had been identified by means of ion-exchange chromatography by establishing their analogy to respective rare-earth elements. But short-lived isotopes will decay before they leave the chromatographic column and will thus distort the chemical picture.

Nature seemed to build an unsurmountable barrier across the way to the second hundred of elements.

MENDELEVIUM

MENDELEVIUM – Synthesized, radioactive metallic element of the actinide series, symbol Md, atomic number 101, relative atomic mass 258. It was first produced by bombardment of Es-253 with helium nuclei. Its longest-lived isotope, Md-258, has a half-life of about two months. The element is chemically similar to thulium. It was named by the US physicists at the University of California at Berkeley who first synthesized it in 1955 after the Russian chemist Mendeleev, who in 1869 devised the basis for the periodic table of the elements.

Scientists made great progress having had synthesized element 100 whose name at last honoured Enrico Fermi who had been the first to start on the quest for transuranium elements.

But beyond fermium one could distinctly see the outlines of a great danger posed by the main enemy of the researchers working with transuranium elements, namely spontaneous fission. According to calculations, the isotopes with $Z = 100$ should have very short half-lives owing to this mechanism of radioactive transformations. Successful synthesis of einsteinium and fermium in high intensity neutron fluxes at first encouraged researchers. But theorists claimed that there was no possibility of advance beyond fermium since its half-life with respect to spontaneous fission was too short. A nucleus of element 100 will decay into two fragments before it has time to emit beta particles.

But still element 101 proved to be the last element that was synthesized in the classical way involving bombardment with alpha particles. By 1955 Seaborg and his group had accumulated about a billion atoms of einsteinium. This infinitesimal amount of einsteinium was very carefully applied to a gold foil whose cost was fantastically small in comparison with that of einsteinium. The target was bombarded with alpha particles. Scientists thought that the nuclear reaction $(, n)$ $^{256}101$ would occur. Owing to recoil effect, the atoms of element 101 penetrated into the gold foil. After bombardment, the foil was dissolved and the solution was analysed in a chromatographic column. The critical thing was to establish when the fraction containing element 101 left the column and to detect spontaneous fission events.

Only five (!) spontaneous fission events were recorded in the first experiment. But that was enough to identify an isotope of element 101. Later its half-life was found to be three hours and its mass number was 256. The half-life was unexpectedly long and contributed to successful synthesis of this new element. It was named mendelevium (Md) in honour of the great Russian chemist D. Mendeleev who had been the first to use the periodic system for predicting the properties of unknown chemical elements. Thus said the discoverers of mendelevium.

Later, when the symbol Md was permanently settled in box 101 of the periodic table they described their discovery in colourful details. A gloomy feeling dominated in the group, they told. Several careful experiments were performed in an attempt to synthesize and identify element 101, all to no avail. At last, the final decisive experiment was prepared and a success could be expected. At best, they hoped to detect one or two atoms of the stubbornly elusive element 101. Holding their breaths, scientists watched the instrument recording spontaneous fission. An hour had passed; the night was almost over; waiting seemed unending.

Suddenly, the pen of an automatic recorder jerked to the mid-scale and returned back tracing a thin red line. Such a burst of ionization had never been observed in the studies of radioactive materials. Probably, this was a signal of expected fission. After about an hour another signal was recorded. Now researchers were sure that two atoms of element 101 had decayed and it could be added to the list of chemical elements.

Interestingly, the instrument recording fission events was connected to a fire alarm and element 101 each time announced its birth by an ear-splitting ringing.

Twelve years later, mendelevium was found to have a longer-lived isotope with a half-life of two months (mendelevium-258). Its existence made possible a detailed study of the chemical properties of mendelevium. The discovery of mendelevium brought to life a new field of radiochemical studies, namely chemistry of single atoms, with its own specialized techniques. It played a decisive part in chemical studies of successive transuranium elements. The synthesis of mendelevium was a watershed in the history of transuranium elements. All formerly used synthesis techniques were no longer applicable since mendelevium could not be accumulated in amounts sufficient to make a target. Theorists visualized the region beyond element 101 as a country populated with ghosts and inaccessible to explorers; it was clear that the following transuranium elements could exist only for seconds or fractions of a second.

Even if they could be obtained, to study their properties seemed an extremely difficult or just impossible task.

But how to obtain them? What nuclear reactions are suitable for that? Fortunately, by the end of the fifties there was a definite answer to this question; multiply charged ions of the light elements (carbon, oxygen, neon, argon) were to be used as bombarding particles. Then the targets could be made from conventional transuranium elements, namely, plutonium, americium and curium and the problem with the target was resolved. Of course, it would be better to have "naked" nuclei for bombardment (such as the alpha particle which is the nucleus of helium) but it was hardly possible to "Skin" the atoms completely. The multiply charged ions had to be accelerated to high energies sufficient for their entering into nuclear reactions. Therefore, new powerful accelerators, were needed. When they had been built a new breakthrough period started in the history of transuranium elements. However, when we talk about discoveries here the word will have a somewhat different sense than in our previous discussions.

NOBELIUM

NOBELIUM – Synthesized, radioactive, metallic element of the actinide series, symbol No, atomic number 102, relative atomic mass 259. It is synthesized by bombarding curium with carbon nuclei.

It was named in 1957 for the Nobel Institute in Stockholm, Sweden, where it was claimed to have been first synthesized. Later evaluations determined that this was in fact not so, as the successful, 1958 synthesis at the University of California at Berkeley produced a different set of data. The name was not, however, challenged.

Yes, element 102 still has no name attached to it. In most current tables of elements, box 102 is not occupied though the element itself is regarded as being well-studied and long known.

Sometimes one can meet in literature the name nobelium and the symbol No but they are just a result of an experimental error that occurred in 1957. At that time an international group of scientists at the Nobel Institute of Physics in Stockholm for the first time used multiply charged ions for synthesizing a new transuranium element. A target of curium-244 was bombarded with ions of carbon-13. The reaction products allegedly contained the isotopes $^{253}102$ and $^{251}102$ with half-lives of about 10 minutes. The success obtained with mendelevium prompted the group to use ion-exchange chromatography the results of which apparently, evidenced the existence of element 102, too.

The claim proved to be erroneous and the experiments have not been substantiated. A current joke at the time was that the only thing left from nobelium was "No".

In Autumn 1957, a group of Soviet scientists headed by G. Flerov entered the field of syntheses of transuranium elements. At present the laboratory of nuclear reactions of the Joint

Institute for Nuclear Research (Dubna, USSR) occupies a leading position in this field. Flerov and his group bombarded a plutonium target with oxygen ions. But the results did not correspond to those reported by the Stockholm group a year earlier. Meanwhile, a group in Berkeley headed now by Seaborg's student, A. Giorso, also attacked element 102. Their results refuted the Stockholm results, too, but did not agree with the Dubna results.

Thus nobelium gradually was reduced to No. Indeed, the date of discovery of this element can hardly be pinpointed. Flerov's group worked on element 102 in 1963-1966. They synthesized several of its isotopes and estimated their mass numbers and half-lives. This was the first real assessment of the new element and the Dubna group had the right to suggest a name for it; it was joliotium in honour of F. Joliot-Curie. But American scientists did not agree with the name though they confirmed the results of the Dubna group.

The arguments about element 102 started the wave of priority controversies which became especially heated for the next transuranium elements. Currently, nine isotopes of element 102 are known, the longest-lived isotope ^{259}102 has a half-life of about one hour.

LAWRENCIUM

LAWRENCIUM – Synthesized, radioactive, metallic element, the last of the actinide series, symbol Lr, atomic number 103, relative atomic mass 262. Its only known isotope, Lr-257, has a half-life of 4.3 seconds and was originally synthesized at the University of California with boron nuclei. The original symbol Lw, was officially changed in 1963.

The element was named after Ernest Lawrence (1901-1958), the US inventor of the cyclotron particle accelerator.

Here we also cannot give the name of the element. And its date of discovery given in the table of discoveries of elements presented in the Conclusion below is not really reliable.

Giorso and his coworkers started the hunt for the new transuranium element in early 1961. A californium target was bombarded with boron ions. Apparently, they obtained the isotope $^{257}103$ with a half-life of 8s.Oof course, they did not hesitate and named the element lawrencium (Lw) in honour of the inventor of cyclotron E. Lawrence. This symbol can often be found in box 103 of the periodic table.

The same isotope $^{257}103$ was synthesized at the Dubna Institute and its properties proved to be quite different from those reported by the Berkeley group. Therefore, they had to change their view and to assume that, in spring 1961 they synthesized not $^{257}103$ but some other isotope, say $^{258}103$ or $^{259}103$.

The situation was clarified in 1965 when the Dubna group carried out the nuclear reaction ^{243}Am (^{18}O, 5n) $^{256}103$ giving rise to the isotope with a mass number of 256 and determined its parameters. They coincided with those reported by the Berkeley scientists for the product of the nuclear reaction $^{249}Cf(^{11}B, 4n)$ $^{256}103$ three years later. This is why the discovery date of 1961 can be doubted. But no definite conclusion was reached as to when and who had discovered element 103. As with element 102, researchers, had to work with just a few atoms of element 103. At first, they found the mass numbers and the radioactive properties of the isotopes and only later the methods for evaluating their chemical nature were found.

14

Discoveries of elements

At present around 118 elements are known. They had different fates; scientists of many countries devoted much time and effort to find them in nature or synthesize them artificially. Now, after we have reviewed all the facts, data and events in the history of elements we can make some conclusions.

Table 6 - 4 presents the dates of discovery and the names of discoverers for all chemical elements with the exception of the elements that became known in the antiquity and middle ages. The discoverers of almost ninety elements can be named. About fifty scientists were directly involved in the discoveries of stable natural elements, nine scientists discovered natural radioactive elements (though about 25 scientists took part in the discoveries of radio-elements entering into radio-active families).

More scientists were involved in the discoveries of synthesized elements (more than 30). It is not surprising because many experimenters and theorists (both physicists and chemists) as well as technicians are involved in the work on syntheses of transuranium elements, particularly those with large Z values.

In total, about 100 scientists were involved in filling the boxes of the periodic table as we know it now.

Some of them can be said to be record-holders. If we turn again to the elements found in nature, here the record is held by the Swedish chemist C. Scheele who discovered six elements, namely, fluorine, chlorine, manganese, molybdenum, barium and tungsten. In addition, he, jointly with J. Priestley, discovered oxygen.

The silver medal for the discoveries of new elements could be awarded to W. Ramsay who discovered (though with Co-Workers) argon (with Rayleigh), helium (with Crookes), krypton, neon, and xenon (all with Travers). Each of the following scientists discovered four elements in nature: J. Berzelius (Cerium, Selenium, Silicon and Thorium), H. Davy (Potassium, Calcium, Sodium and Magnesium), and P. Lecoq de Boisbaudran (Gallium, Samarium, Gadolinium and Dysprosium). Three elements were discovered by each of the following scientists: M. Klaproth (Titanium, Zirconium and Uranium), and C. Mosander (Lanthanum, Terbium and Erbium).

Table 6 - 4

Element	Date	Discoverers
Hydrogen	1766	Henry Cavendish
Helium	1868	Pierre Jules Cesar Jensen, Joseph Norman Liocduier
Lithium	1817	Johan Arfvedson
Beryllium	1798	Louis Nicolas Vauquelin
Boron	1808	Joseph Gay-Lussac, Louis Thenard.
Carbon		Known from antiquity
Nitrogen	1772	Daniel Rutherford
Oxygen	1774	Joseph Priestley, Carl Scheele
Fluorine	1771	Carl Scheele
Neon	1898	William Ramsay, Morris Travers

Element	Date	Discoverers
Sodium	1807	Humphry Davy
Magnesium	1808	Humphry Davy
Aluminium	1825	Hans Oersted
Silicon	1823	Jons Berzelius
Phosphorus	1669	Hennig Brandt
Sulphur		Known from antiquity
Chlorine	1774	Carl Scheele
Argon	1894	William Ramsay, John Rayleigh
Potassium	1807	Humphry Davy
Calcium	1808	Humphry Davy
Scandium	1879	Lars Nilson
Titanium	1795	Martin Klaproth
Vanadium	1830	Nils Sefstrom
Chromium	1797	Louis Vauquelin
Manganese	1774	Carl Scheele
Iron		Known from antiquity
Cobalt	1735	Georg Brandt
Nickel	1751	Axel Cronstedt
Copper		Known from antiquity
Zinc		Obtained in middle ages
Gallium	1875	P. Lecoq de Boisbaudran
Germanium	1886	Clemens Winkler
Arsenic		Obtained in middle ages
Selenium	1817	Jons Berzelius

Element	Date	Discoverers
Bromine	1826	Antoine Balar
Krypton	1898	William Ramsay, Morris Travers
Rubidium	1861	Robert Bunsen, Gustav Kirchhoff
Strontium	1790	Atair Crawford
Yttrium	1794	Johan Gadolin
Zirconium	1789	Martin Klaproth
Niobium	1801	Charles Hatchet
Molybdenum	1778	Carl Scheele
Technetium	1937	Carlo Perrier, Emilio Segre
Ruthenium	1844	Carlovitch Claus
Rhodium	1804	William Wollaston
Palladium	1803	William Wollaston
Silver		Known from antiquity
Cadmium	1817	Friedrich Stromeyer
Indium	1863	Ferdinand Reich
Tin		Known from antiquity
Antimony		obtained in middle ages
Tellurium	1782	Franz Muller Von Reichenstein
Iodine	1811	Bernard Courtois
Xenon	1898	William Ramsay, Morris Travers
Cesium	1861	Robert Bunsen, Gustav Kirchhoff
Barium	1774	Carl Scheele, Otto Hahn
Lanthanum	1839	Carl Mosander
Cerium	1803	Jons Berzelius, Wilhelm Hisinger, Martin Klaproth
Praseodymium	1885	Carl Auer Von Welsbach

Element	Date	Discoverers
Neodymium	1885	Carl Auer Von Welsbach
Promethium	1945	J. Marinsky, L.Glendenin, C. Coryell
Samarium	1879	Paul Lecoq de Boisbaudran
Europium	1901	Eugene Demarcay
Godolinium	1886	Paul Lecoq de Boisbaudran
Terbium	1843	Carl Mosander
Dysprosium	1886	Paul Lecoq de Boisbaudran
Holmium	1879	Per Cleve
Erbium	1843	Carl Mosander
Thulium	1879	Per Cleve
Ytterbium	1878	Charles Marignac
Lutecium	1907	Georg Urbaine
Hafnium	1923	Georg Hevesi, Dirk Coster
Tantalum	1802	Anders Ekeberg
Tungsten	1781	Carl Scheele
Rhenium	1925	Walter Noddack, Ida Tacke, OHO Berg
Osmium	1804	Smithson Tennant
Iridium	1804	Smithson Tennant
Platinum	1748	Antonio de Ulloa
Gold		Known from antiquity
Mercury		Known from antiquity
Thallium	1861	William Crookes
Lead		Known from antiquity
Bismuth		Obtained in middle ages
Polonium	1898	Marie curie, Pierre curie

Element	Date	Discoverers
Astatine	1940	Dale R. Corson, K. R. Mackenzie, Emilio, Segre.
Radon	1899	Ernest Rutherford, Robert Owens.
Francium	1939	Marguerite Perey
Radium	1898	Marie Curie, Pierre Curie
Actinium	1899	Andre Debierne
Thorium	1828	Jons Berzelius
Protactinium	1918	Otto Hahn, Lise Meitner, Fredrich Soddy, A. Cranston
Uranium	1789	Martin Klaproth
Neptunium	1940	Edwin Mcmillan, Philip Abelson
Plutonium	1940	Glenn Seaborg,
Americium	1945	Glenn Seaborg
Curium	1944	Glenn Seaborg
Berklium	1950	Glenn Seaborg
Californium	1950	Glenn Seaborg
Einsteinium	1952	Albert Giorso, G. Seaborg,
Fermium	1952	Albert Giorso, G. Seaborg,
Mendelevium	1955	Glenn Seaborg
Nobelium	1963-1966	Georgil Flerov
Lawrencium	1961	Albert Giorso

Finally, several scientists discovered two elements each: L. Vauquelin (Beryllium and Chromium), W. Wollaston (Rhodium and Palladium), R. Bunsen and G. Kirchhoff (Rubidium and Cesium), C. Auer Von Welsbach (Praseodymium and Neodymium),

P. Cleve (Holmium and Thulium) and S. Tennant (Osmium and Iridium). This is a somewhat idealized description. When we discussed the discoveries of individual elements, we not once met with a situation when the discoverer could not be named.

As for the natural radioactive elements, the champions here are the Curies who extracted polonium and radium from uranium ore. G. Seaborg took part in the discoveries of eight transuranium elements (from Plutonium to Mendelevium). G. Flerov and his large group from Dubna played a decisive role in reliable syntheses of elements 102-107.

Now let us look at the discoveries of elements in various countries.

The largest number of elements 23–were discovered by the Swedish scientists. They include (in chronological order) Cobalt (1735) Nickel (1751), Fluorine (1771), Chlorine (1774), Manganese (1774), Barium (1774), Molybdenum (1778), Tungsten (1781), Yttrium (1794), Tantalum (1802), Cerium (1803), Lithium (1817), Selenium (1817), Silicon (1823), Thorium (1828), Vanadium (1830), Lanthanum (1839), Terbium (1843), Erbium (1843), Scandium (1879), Holmium (1879) and Thulium (1879). This list contains many rare and rare-earth elements and it is not surprising. In Sweden of the 18th century metallurgy was well developed and new deposits of iron ores were needed. The scientists who searched for them discovered at the same time, or often independently, new minerals which were found to contain unknown elements. Moreover, Swedish chemists accumulated considerable experience in analysing various minerals and ores. Thus, the practical requirements of the industry made Sweden the country whose scientists discovered the greatest number of elements.

The second place was held by Britain. British scientists discovered a total of 20 elements : Hydrogen (1766), Nitrogen (1772), Oxygen (1774), Strontium (1787), Niobium (1801), Palladium (1803) Rhodium (1804), Osmium (1804), Iridium

(1804), Sodium (1807), Potassium (1807), Magnesium (1808), Calcium (1808), Thallium (1861), Argon (1894), Helium (1895), Neon (1898), Krypton (1898), Xenon (1898), Radon (1900). The work of British chemists especially clearly demonstrates the links between the general orientation of research and the discoveries of elements. In Britain, the birthplace of pneumatic chemistry, there were discovered the varieties of air which later proved to be the elementary atmospheric gases, namely, hydrogen, nitrogen and oxygen. More than a hundred years later inert gases were discovered in Britain owing to a favourable situation in this field of research (here an outstanding role was played by one scientist, namely, W. Ramsay). In the early 19th century electro-chemistry made significant advances in Britain, which made it possible for H. Davy to produce free sodium, potassium, magnesium and calcium. The discovery of the four platinum metals was due to the progress of studies of raw platinum in Britain.

The third place is held by France where fifteen elements were discovered : Chromium (1797), Beryllium (1798), Boron (1808), Iodine (1811), Bromine (1826), Gallium (1875), Samarium (1879), Gadolinium (1886), Dysprosium (1886) Radium (1898), Polonium (1898), Actinium (1899), Europium (1901), Lutecium (1907), Francium (1939). It is not surprising that the radioactive elements polonium, radium and actinium were discovered by French scientists. These discoveries proceeded from the pioneering studies of radioactivity conducted in France. A brilliant spectral analyst P. Lecoq de Boisbaudran discovered by means of spectral analysis four new elements – Gallium and three rare-earth elements (Samarium, Gadolinium and Dysprosium). Chromium and Beryllium were discovered by L. Vauquelin who was such a skillful analytical chemist that it would be unjust if he had not given the world at least one new element.

Germany holds the fourth place in the number of discovered elements (10). These include Zirconium (1789), Uranium (1789) Titanium (1795), Cadmium (1817), Cesium (1860), Rubidium (1861), Indium (1863), Germanium (1886), Protactinium (1918), Rhenium (1925). The following three factors greatly contributed to these discoveries : the brilliant skill of the analytical chemist M.Klaproth (Ti, Zr and U), development of spectral analysis (Cs, Rb, and In), and wide-ranging X-ray spectral studies (Re).

Austrian scientists discovered three elements : Tellurium (1782), Praseodymium (1885) and Neodymium (1885). Danish scientists discovered Aluminium (1825) and Hafnium (1923): one element (Ruthenium) was discovered in Russia in 1844. But Russian scientists extracted many newly discovered elements from natural minerals and studied their properties (Platinum metals, Chromium, Strontium), Though for a variety of reasons Russian chemists did not discover many new elements one should not forget that the periodic system of elements was developed by the great Russian chemist D. Mendeleev and this task was much more difficult than to discover a few new elements.

It is not surprising that the overwhelming majority of elements found in nature were discovered in the four countries – Britain, France, Germany and Sweden – in which chemical sciences were highly developed. Scientists of these countries obtained many significant results contributing to the discoveries of new elements.

Another interesting question is the rate of discoveries of elements in various historical periods. Let us start with 1750 (which is about the time when chemical analysis started to develop) and end with 1925 (when the last stable element – Rhenium – was discovered). The data for each 25 year period is given in Table 6 - 5.

The Rate of Discoveries of New Elements Between 1750 and 1925

Years	Discovered elements	Total No. of known elements
Before 1750	16 (C,P,S,Fe, Co, Cu, Zn, As, Ag, Sn,	16
	Sb, Pt, Au, Hg, Pb, Bi)	
1751-1775	8 (H, N, O, F, Cl, Mn, Ni, Ba)	24
1776-1800	10 (Be, Ti, Cr, Y, Zr, Mo, Te, W, U, Sr)	34
1801-1825	18 (Li, B, Na, Mg, Al, Si, K, Ca, Se,	52
	Nb, Rh, Pd, Cd, I, Ce, Ta, Os, Ir)	
1826-1850	7 (V, Br, Ru, La, Tb, Er, Th)	59
1851-1875	5 (Rb, In, Cs, Ti, Ga)	64
1876-1900	19 (He, Ne, Ar, Sc, Ge, Kr, Xe, Pr, Nd, Sm, Gd, Dy, Ho, Tu, Yb, Po, Ra, Ac, Rn)	83
1901-1925	5 (Eu, Lu, Hf, Re, Pa)	88

Table 6-5 demonstrates that two 25 year periods were particularly rich in discoveries of new elements. The first period is from 1801 to 1825 when 18 elements were discovered. This is easy to understand as this period saw a great progress in chemical analysis owing to the work of such outstanding scientists as Klaproth, Berzelius and others. A signficant contribution was made by Davy who introduced the electro-chemical method which immediately yielded several alkali and alkaline-earth metals. The second peak period is explained by the development of spectrometry and radiometry and the advances in the chemistry of rare earths (as can be readily understood when looking at the symbols of the 19 elements discovered in this period). But in the fifty years between these periods only 12 new elements were discovered (1825–1875). The reasons for that are simple. Chemical analysis at this period, so to say, picked the leftovers, that is, the few remaining elements it had the capacity to identify. On the other hand, spectral analysis was still a young science, just testing its strength. The fact that in the first quarter of the 20th century only five elements were discovered does not mean that the capabilities of science were limited; it just demonstrates that the naturally occurring elements have practically all been found.

The above discussion has one weak point which somewhat diminishes its value. It is based on the data given is Table 6-4, particularly, on the dates of discoveries (when they are known at all). But these dates describe different events in the history of elements or, in other words, are of varying significance.

This can be shown with the following simple examples. Take three halogens–Fluorine, Chlorine and Bromine. The date of fluorine discovery is considered to be 1771 when C.Scheele prepared a substance that later proved to be hydrofluoric acid. But it was only fifteen years later that Lavoisier suggested that it contained a new element and he was mistaken, into the bargain, assuming that the acid contained oxygen. It was only in 1810 that Davy and Ampere definitely stated that hydro-fluoric acid was a compound of hydrogen and an unknown element, that is, fluorine. The element was produced in a free form as late as 1886. Generally speaking, each of these dates can be regarded as the date of discovery of fluorine. But the chosen date is 1771 though Scheele did not definitely know what he had discovered.

Chlorine was also discovered by Scheele in the form of deflogisticated muric acid and he did not regard it as a simple substance though he observed precisely the evolution of a free halogen. This fact makes the accepted discovery date for chlorine (1774) better substantiated than that for fluorine for which the simple substance had yet to be extracted. The decisive event in the history of chlorine was the establishment of its elementary nature in 1810 by Davy. And Davy is regarded as the discoverer of sodium, potassium, magnesium and calcium though compounds of these elements had long been known.

On the other hand, iodine is a good example of an element which did not give rise to any controversies. It was discovered in 1811 directly as a simple substance, studied within a short period of time and recognized as a relative of halogens. Thus, we see that three dates of discoveries of related elements (1771, 1774 and 1811) given in Table 6-4 have quite different meanings.

Another example is given by discoveries of three totally unrelated elements–Bromine, Yttrium and Helium. What is the meaning of their dates of discovery in Table 6-4? The date for bromine (1826) corresponds to the extraction of the element in a free form. The date for yttrium corresponds to the preparation of its oxide (1794). Forty years later it became clear that the "yttrium" of Gadolin had in fact been a mixture of rare earths, and a relatively clean yttrium oxide was prepared by Mosander. Thus, in 1794 a mixture of related elements was discovered rather than an individual element. The accepted date of discovery of helium (1868) corresponds to an event which had never before happened in the history of elements. For the first time a conclusion about the existence of a new element was made proceeding from an unknown line in the spectrum of solar prominences rather than from experiments with material terrestrial objects. This element remained a pure hypothesis until it was found on Earth (1895).

Again we see that three discovery dates have different meanings and backgrounds. We can give more examples.

How then to explain this distinct difference in meaning between the dates of discoveries of elements? The answer is that the term "discovery of chemical elements" has no clear definition and is often used in different contexts.

"Discovery of an element must mean not only preparation (extraction) of the element in a free form but also determination of its existence in some compounds with chemical or physical means. Naturally, this definition is applicable only to the discoveries made starting with the second half of the 18th century." It cannot be applied to the earlier historical periods when scientists had no means for studying the composition of the compounds containing unknown elements." We fully agree with the latter part of the above statement but not with the first phrase. It does not differentiate between preparation of a new element in the form of a simple substance and determination of its existence in compounds. But these are quite different things as we have shown when we discussed the different meanings of the dates of discoveries for different elements. Isolation of an element in the form of a simple substance is an important event in its history. Indeed, to obtain sufficient knowledge of the properties of an element the element must be available in a free form. Only then can Scientists study many of its chemical properties (for instance, its reactions with various reactants) and almost all physical properties. Therefore, extraction of an element in a free form should be regarded as a higher stage of discovery and its preparation in the form of a compound as a lower, preliminary stage.

The history of elements evidences that the higher stage of discovery was reached by no means always, that is, the discovery did not always mean that the element was prepared in a free form. Thus, in many cases we cannot consider the discovery of an element as a single event. It is rather a more or less protracted process. Table 6–4 gives only one date in the history of an element and thus, in a way, ignores the history itself. Some dates in the table even correspond to indirect determination of the existence of a new element (for the elements that at first were discovered radiometrically or spectroscopically but were not extracted materially at all.)

We can classify chemical elements into two groups according to the methods of their discovery : elements found in nature and synthesized elements. We shall not consider those elements in the first group to which the concept of discovery is inapplicable, that is, those known from antiquity or the middle ages.

Then we see that a large part of the first group consists of the elements that were first obtained in compounds (Li, Be, F, Sc, Ti, V, Rb, Y, Zr, Nb, Mo, La, Ce, Pr, Nd, Sm, Eu, Gd, Tb, Dy, Ho, Er, Tu, Yb, Lu, Hf, Ta, W, Re, Po, Fr, Ra, Ac, Pa, Th, U). Out of these 36 elements the existence of ten elements (Rb, Sm, Eu, Dy, Ho, Tu, Yb, Lu, Hf, Re) was first determined spectros-copically and of five elements (Po, Fr, Ra, Ac, Pa) radiometrically. Some of the above elements can be placed into this list only conditionally though.

Almost as many elements (40) were obtained in a free state without previous identification in compounds (H, He, B,N,O, Ne, Na, Mg, Al, Si, P,Cl, Ar, K, Ca, Cr, Mn, Co, Ni, Ga, Ge, Se, Br, Kr, Sr, Ru, Rh, Pd, Cd, In, Te, I, Xe, Cs, Ba, Os, Ir, Pt, Ti, Rn). The existence of eight elements (He, Ne, Ar, Kr, Xe, Cs, In, Tl) was first established spectroscopically and radon was first found radiometrically.

Thus, the history of chemical elements is far from being complete. It needs new studies and reassessment of old data; it is still capable of unexpected findings which could make us review seemingly indisputable opinions. Strange as it may seem in the world literature, there is still no fundamental study giving a detailed and comprehensive analysis of the history of discoveries of elements.

This book is just an attempt to present a general outline of this history. At the end, we shall talk briefly on a subject which has a direct bearing on the history of elements, namely, the false (erroneous) discoveries of elements. Nobody has yet attempted to compile an exhaustive list of mistakes in the history of elements as it is a very difficult task (for many reasons). We shall just give here the names of about a hundred erroneously discovered elements (giving the dates and the names of discoverers) and briefly analyse the causes of mistakes. More than a half of them were made in the studies of rare-earth elements (there were perhaps twice as many mistakes in these studies but in many cases the "discovered" elements were not named but were just designated with Latin or Greek letters). Here is a list of these false elements in alphabetical order: austrium (1886, E. Linnemann), berzelium (1903, C. Baskerville), Carolinium (1900, C. Baskerville), Celtium (1911, G. Urbain), Columbium (1879, G. Smith), damarium (1896, K. Lauer, P. Antsch), decipium (1878, M. Delafontaine), demonium (1894, H. Rowland), denebium (1916, G. Eder), donarium (1851, C. Bergmann), dubhium (1916, G. Eder), eurosamarium (1917, G. Eder), euxenium earth I,II (1901, K. Hoffmann, W. Prandtl), glaukodymium (1897, K. Khrushchev), incognitum and ionium (1905, W. Crookes), Junonium (1811, T. Thomson), Kosmium (1896, B. Kosmann), Lucium (1896, P. Barriere), Masrium (1892, H. Richmond), metacerium (1895, B. Brauner), monium or victorium (1898, W. Crookes), mosandrium (1877,G. Smith), neokosmium (1896, B. Kosmann), Philippium (1878, M. Delafontaine), rogerium (1879, G. Smith), Russium (1887, K. Khrushchev), Vestium (1818, L. Gilbert), Wasium (1862, G. Bohr), Welsium (1920, G. Bahr). Other erroneously discovered rare earths remained nameless.

Many false discoveries are connected with the search for elements 43, 61, 85 and 87 which scientists long and unsuccessfully tried to find in nature, mainly, in the first four decades of this century. Here are a few examples: alabamium (1931,F. Allison et al.) alcalinium (1926, F. Loring, G.Druce), dacinum (1937, R. de Separet), Florentium (1926, L. Rolla, L. Fernandes), helvetium or anglohelvetium (1940, W. Minder ; 1942, A. Leigh-Smith), illinium (1926, D. Harris et al.), Leptine (1943, K.Martin), masurium (1925,W. Noddack, I. Tacke, O. Berg), moldavium (1937, H. Hulubei), nipponium (1908, M. Ogawa), Russium (1925, D. Dobroserdov), virginium (1930, F. Allison et al.). In the mid-thirties erroneous reports on discoveries of transuranium elements appeared, too (for instance, ausonium, hesperium, bohemium, sequanium).

A large number of false discoveries were made in the studies of ores and minerals with complex compositions, particularly crude platinum. For Instance, the following erroneously discovered elements were reported: amarillium (1903, W. Curtis), Canadium (1911, A. French), davyum (1877, S.Kern), josefinium (1903, discoverer unknown), Oudalium (1879, A. Guyard), Pluranium, Polinium and ruthenium (1829, G.Osann), Vestium(1808, J. Sniadecki). Discoveries of new platinum metals which remained unnamed were reported by F. Genth (1853), C. Chandler (1862) and T. Wilm (1883). This list, of course, is far from being complete.

The studies of columbites and minerals of cobalt, zirconium and nickel also led to false discoveries, for instance, dianium (1860, F. Kobell), gnomium (1889, G. Kruss, F. Schmidt), idunium (1884, H. Websky), ilmenium (1846, R. Hermann), jargonium (1869, H. Sorby), neptunium (1850, R. Hermann), nigrium (1869, A. Church), niccolanum (1803, T. Richter), norwegium (1879, T. Dahl,) norium (1845, A. Svanberg), Ostran (1825, A.Breithaupt), Pelopium (1846, H. Rose), Vestium or Sirium (1818, L. Von West), Vodanium (1818, V. Lampadius).

These are four large groups of false discoveries. Apart from them, many single erroneous discoveries of a chance character are known to history, for instance, austrium (1889, B. Brauner), actinium (1881, T. Phipson), Crodonium (1820, I. Trommsdorf), donium (1836, A. Richardson), eka-tellurium (1889, A. Grunwald), etherion (1898, C.Brush), lavoesium (1877, G. Prat), metaargon (1898, W. Ramsay, M. Travers), Oceanium (1923, A.Scott), panchromium or erytronium (1801, A.del Rio), treenium (1836, G.Boase), Vesbium (1879, A.Scacchi).

The above names are sometimes repeated (austrium, vestium) or coincide with the names of real elements (actinium, ruthenium). These are chance coincidences. The elements denoted with an asterisk are of special interest. In their cases there are grounds to think that the analysed specimens indeed contained unknown elements which could not be identified. Here it would be more correct to speak about unrecognized, rather than false, elements. For instance, amarillium and davyum could, apparently, be regarded as possible precursors of rhenium and nipponium as a precursor of hafnium.

All these erroneously discovered elements were found in experiments that were performed more or less correctly but whose results were, as a rule, misinterpreted. However, in old chemical and physical literature one can meet names of elements which were never discovered. These are so-called hypothetical elements whose existence was only postulated for explaining some processes or assumed on the basis of indirect evidence (for instance, coronium, nebulium, asterium, arconium and protofluorine, whose existence was assumed in various cosmic bodies). In fact, they have no bearing on the history of chemical elements.

www.ingramcontent.com/pod-product-compliance
Lightning Source LLC
Chambersburg PA
CBHW052308220526
45472CB00001B/20